T0225060

SpringerBriefs in Mathematical Physics

Volume 42

SpringerBriefs are characterized in general by their size (50–125 pages) and fast production time (2–3 months compared to 6 months for a monograph).

Briefs are available in print but are intended as a primarily electronic publication to be included in Springer's e-book package.

Typical works might include:

- An extended survey of a field
- A link between new research papers published in journal articles
- A presentation of core concepts that doctoral students must understand in order to make independent contributions
- Lecture notes making a specialist topic accessible for non-specialist readers.

SpringerBriefs in Mathematical Physics showcase, in a compact format, topics of current relevance in the field of mathematical physics. Published titles will encompass all areas of theoretical and mathematical physics. This series is intended for mathematicians, physicists, and other scientists, as well as doctoral students in related areas.

Editorial Board

- Nathanaël Berestycki (University of Cambridge, UK)
- Mihalis Dafermos (University of Cambridge, UK / Princeton University, US)
- Atsuo Kuniba (University of Tokyo, Japan)
- Matilde Marcolli (CALTECH, US)
- Bruno Nachtergaele (UC Davis, US)
- Hal Tasaki (Gakushuin University, Japan)
- 50 – 125 published pages, including all tables, figures, and references
- Softcover binding
- Copyright to remain in author's name
- Versions in print, eBook, and MyCopy

More information about this series at http://www.springer.com/series/11953

Hidehtio Nagao · Yasuhiko Yamada

Padé Methods for Painlevé Equations

 Springer

Hidehtio Nagao
Natural Sciences Division
National Institute of Technology
Akashi College
Akashi, Japan

Yasuhiko Yamada
Department of Mathematics
Kobe University
Kobe, Japan

ISSN 2197-1757   ISSN 2197-1765 (electronic)
SpringerBriefs in Mathematical Physics
ISBN 978-981-16-2997-6   ISBN 978-981-16-2998-3 (eBook)
https://doi.org/10.1007/978-981-16-2998-3

This Springer imprint is published by the registered company Springer Nature Singapore Pte Ltd.
The registered company address is: 152 Beach Road, #21-01/04 Gateway East, Singapore 189721,
Singapore

Preface

The isomonodromic deformation equations such as Painlevé and Garnier systems are a very important class of nonlinear differential equations in mathematics and mathematical physics. In parallel with the numerous breakthroughs in discrete integrable systems (see a recent well-organized text book [14]), the study of the isomonodromic systems has also made much progress, particularly in the discrete (difference) cases. Today, we can access isomonodromic equations from various approaches: Painlevé test/Painlevé property, reduction from integrable hierarchies, Lax formulation, theory of orthogonal polynomials, affine Weyl group symmetries, algebraic geometry, cluster algebras, and so on. Among them, the Padé method we will explain in this monograph provides a very simple approach in both continuous and discrete cases.

For a given function $\psi(x)$, the Padé approximation (or interpolation) [2] supplies polynomials $P(x)$, $Q(x)$ as approximants such as $\psi(x) \sim \frac{P(x)}{Q(x)}$. The basic idea of the Padé method is to study the linear differential (or difference) equations satisfied by $P(x)$ and $\psi(x)Q(x)$. Choosing the approximation problem suitably, the linear equations give the Lax pair for some isomonodromic equations. In fact, the close relation between the isomonodromic equations and Padé approximations has been known classically, and the main advantage of this method is its simplicity. Once we set up a suitable Padé approximation/interpolation problem, we can obtain various explicit and *exact* (not approximate) results on the corresponding isomonodromic equations together with their Lax formulations and special solutions. Thus the method offers a very effective and instructive approach to continuous/discrete isomonodromic equations.

The aim of this monograph is to explore a systematic and pedagogical study of the Padé method for both the continuous and discrete cases. In Chap. 1, we explain how the interesting special equations and special functions arise from the Padé problem through a toy example. In the process we will naturally encounter several key concepts on linear differential equations. In Chap. 2, we study the sixth Painlevé equation P_{VI} (the most generic one among the six classical Painlevé equations) through a certain Padé approximation. Again, all the fundamental concepts and equations will arise naturally from the Padé problem. There exists a very nice q-difference analog of the P_{VI} equation: the q-P_{VI} by Jimbo and Sakai. In Chap. 3, we study certain Padé approximation problems related to q-P_{VI} together with its multivariable generalization (the q-Garnier system). The Padé interpolations are a more natural setting in which to approach the discrete Painlevé equations, and there are several kinds of Padé interpolation depending on the "grids" (the choice of the interpolating points: x_s). In Chap. 4, we study the q-Painlevé/Garnier systems using the Padé interpolation on a q-grid: $x_s = q^s$. In Chap. 5, the Padé interpolation on a q-quadratic-grid: $x_s = q^s + cq^{-s}$ is studied, and the result is related to a version of q-Garnier systems. Finally, Chap. 6 is devoted to an introduction to certain multicomponent Padé approximations/interpolations.

As mentioned above, the relation between the isomonodromic equations and Padé approximations has been known classically. However, its systematic exploration including discrete cases was started rather recently. Many colleagues have contributed to make the project come to life, through their excellent work and valuable suggestions. We would in particular like to thank (in alphabetical order) Yusuke Ikawa, Kenji Kajiwara, Toshiyuki Mano, Tetsu Masuda, Masatoshi Noumi, Yasuhiro Ohta, Kanam Park, V. P. Spiridonov, Takao Suzuki, Teruhisa Tsuda, Satoshi Tsujimoto, and A. S. Zhedanov. The contents of this book grew out of the series of lectures by the second-named author at the University of Sydney in 2016. He sincerely thanks Prof. Nalini Joshi for providing such a nice opportunity, and the audiences for their interest. We would like to thank the referees for carefully reading the manuscript and giving valuable comments and suggestions.

Last but not least, we sincerely thank Mr. M. Nakamura of Springer Japan for his patient encouragement.

Akashi, Japan   Hidehito Nagao
Kobe, Japan   Yasuhiko Yamada

Acknowledgements H. Nagao is supported by JSPS KAKENHI Grant Number 19K14579, and Y. Yamada is supported by JSPS KAKENHI Grant Number 26287018.

Contents

Chapter 1
Padé Approximation and Differential Equation

Abstract The Padé approximation gives rational functions $\frac{P(x)}{Q(x)}$ which approximate a given function $\psi(x)$. From the approximants $P(x)$, $Q(x)$, one can obtain interesting linear differential equations. We explain the derivation of such equations in a simple example.

1.1 Linear Differential Equations

Consider a 2nd-order linear differential equation[1] for an unknown function $y = y(x)$:

$$y'' + a(x)y' + b(x)y = 0. \quad ('= \frac{d}{dx}) \tag{1.1}$$

It is enough to get two linearly independent solutions (*fundamental solutions*) to give a general solution; however, finding the fundamental solutions is not so easy in general. Conversely, the construction of the equation from its fundamental solutions is rather easy. Let y_1, y_2 be fundamental solutions, then the differential equation can be written as the *Wronskian* determinant

$$W(y, y_1, y_2) = \begin{vmatrix} y & y' & y'' \\ y_1 & y_1' & y_1'' \\ y_2 & y_2' & y_2'' \end{vmatrix} = 0. \tag{1.2}$$

Example

If we know the fundamental solutions y_1, y_2 of the form

---

[1] Here we will summarize briefly some necessary facts on the linear differential equations. For more explanations, see e.g. [17, 67].

H. Nagao and Y. Yamada, *Padé Methods for Painlevé Equations*,
SpringerBriefs in Mathematical Physics,
https://doi.org/10.1007/978-981-16-2998-3_1

$$y_i = (x - c)^{\rho_i}\{1 + O(x - c)\} \quad (\rho_1 \neq \rho_2), \quad x \to c \tag{1.3}$$

then the corresponding differential equation takes the form

$$y'' + \left\{\frac{1 - \rho_1 - \rho_2}{x - c} + O(1)\right\}y' + \left\{\frac{\rho_1\rho_2}{(x - c)^2} + O(\frac{1}{x - c})\right\}y = 0. \tag{1.4}$$

The form of the solutions (1.3) or the form of the Eq. (1.4) can be used as an easy definition that $x = c$ is a *regular singular point* with *exponents* ρ_1, ρ_2. In the case of $c = \infty$, the point $x = \infty$ is regular singular if the point $w = 0$ is regular singular under the change of variable $x = w^{-1}$. Explicitly it can be characterized by

$$y_i = x^{-\rho_i}\left\{1 + O(\frac{1}{x})\right\}, \quad x \to \infty \tag{1.5}$$

or

$$y'' + \left\{\frac{1 + \rho_1 + \rho_2}{x} + O(1)\right\}y' + \left\{\frac{\rho_1\rho_2}{x^2} + O(\frac{1}{x})\right\}y = 0. \tag{1.6}$$

It is known that when the coefficients $a(x)$, $b(x)$ in (1.1) are holomorphic at some point, then the solutions are also holomorphic there. In other words, the singular points of the solutions are completely determined by the singularities of the coefficients $a(x)$, $b(x)$.[2] If all the singular points are regular singular, such an equation is called *Fuchsian*. The general form of the 2nd-order Fuchsian differential equation with N regular singular points $u_1, \ldots, u_{N-1}, u_N = \infty$ is given by (1.1) with

$$a(x) = \sum_{i=1}^{N-1}\frac{a_i}{x - u_i}, \quad b(x) = \sum_{i=1}^{N-1}\left\{\frac{b_i}{(x - u_i)^2} + \frac{c_i}{x - u_i}\right\}. \tag{1.7}$$

The exponents (α_i, β_i) at each singular point u_i are the basic data of the Fuchsian differential equation. It is represented by the following table called the *Riemann scheme*:

$$\begin{array}{|cccc|}
\hline
x = u_1 & u_2 & \cdots & u_{N-1} \ \infty \\
\hline
\alpha_1 & \alpha_2 & \cdots & \alpha_{N-1} \ \alpha_N \\
\beta_1 & \beta_2 & \cdots & \beta_{N-1} \ \beta_N \\
\hline
\end{array} \qquad . \tag{1.8}$$

Since $a_i = 1 - \alpha_i - \beta_i$ and $\sum_{i=1}^{N-1}(1 - \alpha_i - \beta_i) = 1 + \alpha_N + \beta_N$, we have

---

[2] This is a crucial property of linear differential equations. For the nonlinear equations, the solution may be singular even if the coefficients of the equation are holomorphic.

$$\sum_{i=1}^{N}(\alpha_i + \beta_i) = N - 2. \tag{1.9}$$

This kind of relation among the exponents is called the *Fuchs relation*.

We remark that when $\alpha_i - \beta_i \in \mathbb{Z}$ a certain special situation may occur (see "Non-logarithmic condition" in Sect. 2.1).

1.2 A Toy Example of Padé Approximation

The *Padé approximations* are approximations of a given function by a rational function. A typical setting is as follows.

For a given function $\psi(x)$ holomorphic around $x = 0$, find polynomials $P(x)$, $Q(x)$ of degree m, n respectively, such that

$$\psi(x) = \frac{P(x)}{Q(x)} + O(x^{m+n+1}). \tag{1.10}$$

Namely, the ith derivative of $\psi(x) - \frac{P(x)}{Q(x)}$, or equivalently those of $P(x) - \psi(x)Q(x)$, vanishes at $x = 0$ for $i = 0, 1, \ldots, m + n$.

The condition (1.10) gives $m + n + 1$ linear equations for the $(m + 1) + (n + 1)$ coefficients of the polynomials $P(x)$, $Q(x)$, and hence, it generically determines the polynomials up to a common normalization factor. In the following, we will see that the Padé approximations give interesting differential (or difference) equations.

As a toy example, we consider the case

$$\psi(x) = (1 - x)^\alpha, \tag{1.11}$$

where α is a complex parameter.

Proposition 1.1 *The 2nd-order linear differential equation for y, y', y'' satisfied by $y = P(x)$ and $y = \psi(x)Q(x)$ is explicitly given by*

$$y'' - \left\{\frac{m + n}{x} + \frac{\alpha - 1}{x - 1}\right\}y' + \frac{m(n + \alpha)}{x(x - 1)}y = 0. \tag{1.12}$$

Proof The equation is obtained as the following Wronskian:

$$\begin{vmatrix} y & y' & y'' \\ \mathbf{u} & \mathbf{u}' & \mathbf{u}'' \end{vmatrix} = |\mathbf{u}, \mathbf{u}'|y'' - |\mathbf{u}, \mathbf{u}''|y' + |\mathbf{u}', \mathbf{u}''|y = 0, \qquad (1.13)$$

where $\mathbf{u} = \begin{bmatrix} P \\ \psi Q \end{bmatrix}$. The computation of the coefficients goes as follows.

Computation of $|\mathbf{u}, \mathbf{u}'|$. Noting that

$$\frac{\psi'}{\psi} = \frac{-\alpha}{1-x}, \qquad (1.14)$$

we have

$$|\mathbf{u}, \mathbf{u}'| = \psi \begin{vmatrix} P & P' \\ Q & \dfrac{\psi'}{\psi}Q + Q' \end{vmatrix} = \psi \begin{vmatrix} x^{[m]} & x^{[m-1]} \\ x^{[n]} & \dfrac{x^{[n]}}{1-x} \end{vmatrix} = \psi \frac{x^{[m+n]}}{1-x}, \qquad (1.15)$$

where $x^{[k]}$ stands for a polynomial in x of degree at most k. On the other hand, due to the condition (1.10), we see that

$$|\mathbf{u}, \mathbf{u}'| = \begin{vmatrix} P & P' \\ \psi Q - P & (\psi Q - P)' \end{vmatrix} = O(x^{m+n}) \quad (x \to 0). \qquad (1.16)$$

Hence, we have

$$|\mathbf{u}, \mathbf{u}'| = \psi \frac{Cx^{m+n}}{1-x} = Cx^{m+n}(1-x)^{\alpha-1}, \qquad (1.17)$$

where C is a constant.

Computation of $|\mathbf{u}, \mathbf{u}''|$. Taking a derivative of (1.17), we have

$$|\mathbf{u}, \mathbf{u}''| = |\mathbf{u}, \mathbf{u}'|' = Cx^{m+n}(1-x)^{\alpha-1} \left\{ \frac{m+n}{x} + \frac{1-\alpha}{1-x} \right\}. \qquad (1.18)$$

Computation of $|\mathbf{u}', \mathbf{u}''|$. Using the relation

$$\frac{\psi''}{\psi} = \frac{\alpha(\alpha-1)}{(1-x)^2}, \qquad (1.19)$$

we have

$$|\mathbf{u}', \mathbf{u}''| = \psi \begin{vmatrix} P' & P'' \\ \dfrac{\psi'}{\psi}Q + Q' & \dfrac{\psi''}{\psi}Q + 2\dfrac{\psi'}{\psi}Q' + Q'' \end{vmatrix} \tag{1.20}$$

$$= \psi \begin{vmatrix} x^{[m-1]} & x^{[m-2]} \\ \dfrac{x^{[n]}}{1-x} & \dfrac{x^{[n]}}{(1-x)^2} \end{vmatrix} = \psi \dfrac{x^{[m+n-1]}}{(1-x)^2}. \tag{1.21}$$

Again, due to the condition (1.10), this is divisible by x^{m+n-1}. Hence

$$|\mathbf{u}', \mathbf{u}''| = C' x^{m+n-1}(1-x)^{\alpha-2}, \tag{1.22}$$

where C' is another constant.

Combining (1.17), (1.18), (1.22), we obtain

$$y'' - \left\{ \frac{m+n}{x} + \frac{\alpha-1}{x-1} \right\} y' + \frac{d}{x(x-1)} y = 0. \tag{1.23}$$

Finally, the constant $d = C'/C$ can be determined as $d = m(n+\alpha)$ since the polynomial $P(x)$ of degree m is a solution. $\qquad\square$

Simple derivation from singularity data

One can derive the Eq. (1.12) from the Riemann scheme

$$\begin{array}{|ccc|} \hline x=0 & 1 & \infty \\ \hline 0 & 0 & -m \\ m+n+1 & \alpha & -\alpha-n \\ \hline \end{array}. \tag{1.24}$$

Moreover, this data can be obtained directly from the singularity structure of the solutions $P(x)$, $\psi Q(x)$. Note that the exponents at $x = 0$ are 0 and $m + n + 1$ since we have fundamental solutions $P(x) = O(x^0)$ and $P(x) - \psi(x)Q(x) = O(x^{m+n+1})$ around $x = 0$.

From the example discussed above, we observe the following lessons. (i) The differential equation can be computed without knowing the explicit form of the polynomials $P(x)$, $Q(x)$. In fact, one can directly obtain the differential equation only by looking at the singularity structure of the solutions $P(x)$, $\psi(x)Q(x)$. (ii) Regardless of the degrees of polynomials $P(x)$, $Q(x)$, the coefficients of the differential equation are given by rational functions of small degree. This is because the Wronskian (1.13) is divisible by x^{m+n-1} thanks to the approximation condition

(1.10). These are general features of the differential equations arising from the Padé approximation.

1.3 Contiguity Relations

Besides the variable x, the polynomials $P(x)$, $Q(x)$ depend on the other parameters α, m, n. Here we will consider the discrete deformation w.r.t. the parameter α.

Proposition 1.2 *For the shift $T^{\pm} : \alpha \mapsto \alpha \pm 1$, we have the differential/difference equations satisfied by $y = P(x)$ and $y = \psi(x)Q(x)$ as follows:*

$$xy' + (\alpha + n)T^{-1}(y) - (\alpha + n)y = 0, \tag{1.25}$$

and

$$(mx - m + \alpha)y + (m - \alpha)T(y) + x(1 - x)y' = 0. \tag{1.26}$$

Proof These equations can be obtained from

$$\begin{vmatrix} y & y' & T^{\pm}(y) \\ u & u' & T^{\pm}(u) \end{vmatrix} = |u, u'|T^{\pm}(y) - |u, T^{\pm}(u)|y' + |u', T^{\pm}(u)|y = 0, \tag{1.27}$$

where $u = \begin{bmatrix} P \\ \psi Q \end{bmatrix}$ as before.

Using the relation

$$T^{-1}(\psi(x)) = (1 - x)^{\alpha - 1} = (1 - x)^{-1}\psi(x), \tag{1.28}$$

and counting the degree and vanishing order in x at $x = 0$ and $x = \infty$, we have

$$|u, T^{-1}(u)| = \frac{\psi(x)}{1 - x}C_1 x^{m+n+1}, \tag{1.29}$$

$$|u', T^{-1}(u)| = \frac{\psi(x)}{1 - x}C_2 x^{m+n}, \tag{1.30}$$

where C_1, C_2 are some constants. Together with (1.17), we have

$$CT^{-1}(y) - C_2 xy' + C_2 y = 0. \tag{1.31}$$

The ratios of constants C, C_1, C_2 can be determined by the condition that the equation has solutions of the form $y = P = 1 + O(x)$ $(x \to 0)$ and $y = \psi Q = O(x^{\alpha + m})$ $(x \to \infty)$, as in (1.25).

Similarly, from $T(\psi(x)) = (1 - x)\psi(x)$, we see

$$|\mathbf{u}, T(\mathbf{u})| = \psi(x)C_3 x^{m+n+1},$$

$$|\mathbf{u}', T(\mathbf{u})| = \frac{\psi(x)}{1-x}(C_4 x + C_5)x^{m+n}. \tag{1.32}$$

Hence we have

$$(C_4 x + C_5)y - CT(y) + C_3 x(1-x)y' = 0. \tag{1.33}$$

Again, the ratios of constants C, C_3, C_4, C_5 can be determined from the existence of solutions of the form $y = 1 + O(x)$ $(x \to 0)$, $y = O(x^m)$ $(x \to \infty)$ and $y = O((1-x)^\alpha)$ $(x \to 1)$, and we have (1.26). □

Contiguity relations

Equations (1.25), (1.26) describing the α-parameter dependence are called *contiguity relations*. Eliminating $T(y)$, $T^{-1}(y)$ from these two equations, one can recover the differential equation (1.23). Furthermore, eliminating y', one can also derive the following difference equation w.r.t. α:

$$(n+\alpha)(x-1)T^{-1}(y) + \left\{(m-n)(x-1) + (2-x)\alpha\right\}y + (m-\alpha)T(y) = 0. \tag{1.34}$$

In this sense, the contiguity relations (1.25), (1.26) are more fundamental than (1.23) and (1.34).

1.4 Explicit Solutions

Though we do not need the explicit form of $P(x)$, $Q(x)$ to derive the differential (and contiguity) equations, the information about the explicit form also plays an important role in our story. In the case of the toy example, they can be determined by solving the differential equation (1.12), and given as

$$\begin{aligned}P(x) &= {}_2F_1(-m, -\alpha - n, -m - n; x), \\ Q(x) &= {}_2F_1(-n, \alpha - m, -m - n; x),\end{aligned} \tag{1.35}$$

where ${}_2F_1$ is the *Gauss hypergeometric function*

$$_2F_1(a, b, c; x) = 1 + \frac{ab}{c1}x + \frac{a(a+1)b(b+1)}{c(c+1)2!}x^2 + \cdots. \tag{1.36}$$

Note that ${}_2F_1(a, b, c; x)$ is polynomial (terminating) if $a \in \mathbb{Z}_{\leq 0}$ or $b \in \mathbb{Z}_{\leq 0}$ and $c \notin \mathbb{Z}_{\leq 0}$. The function ${}_2F_1(a, b, c; x)$ is a solution of the following equation:

$$x(1-x)y'' + \{c - (a+b+1)x\}y' - aby = 0. \tag{1.37}$$

This equation (*Gauss hypergeometric equation*) is characterized by the Riemann scheme

$$\begin{array}{|ccc|}\hline x=0 & 1 & \infty \\\hline 0 & 0 & a \\ 1-c & c-a-b & b \\\hline\end{array}. \tag{1.38}$$

The Eq. (1.12) is a special case of this with two exponents in $\mathbb{Z}$.

Derivation from Euler transformation

The Padé approximation for $\psi(x) = (1-x)^\alpha$ can be obtained from the *Euler transformation formula*:

$$(1-x)^{a+b-c}{}_2F_1(a,b,c;x) = {}_2F_1(c-a,c-b,c;x), \tag{1.39}$$

which follows by noting the change of exponents

$$(1-x)^{a+b-c} \times \begin{array}{|ccc|}\hline x=0 & 1 & \infty \\\hline 0 & 0 & a \\ 1-c & c-a-b & b \\\hline\end{array} = \begin{array}{|ccc|}\hline x=0 & 1 & \infty \\\hline 0 & 0 & c-a \\ 1-c & -c+a+b & c-b \\\hline\end{array}, \tag{1.40}$$

and consider the holomorphic solution around $x = 0$ which is generically unique up to a normalization.

Naively, a specialization $(a, b, c) = (-m, -\alpha - n, -m - n)$ of the formula (1.39) seems to give the approximation relation (1.10), but the error terms are missing. The correct result can be obtained by putting $(a, b, c) = (-m, -\alpha - (n + \epsilon), -m - (n + \epsilon))$ and taking the limit $\epsilon \to 0$. This is exactly the method H. Padé used in [48] to obtain the approximation of the power function $(1 - x)^\alpha$ as ratios of the terminating hypergeometric functions.

In a similar way, starting from the function $\psi(x) = e^x$, one obtains the differential equation

$$xy'' - (x + m + n)y' + my = 0. \tag{1.41}$$

The corresponding Padé interpolants are $P(x) = {}_1F_1(-m, -m - n; x)$ and $Q(x) = {}_1F_1(-n, -m - n, x)$, where ${}_1F_1$ is the *confluent hypergeometric function*

$$_1F_1(\alpha, \gamma; x) = 1 + \frac{\alpha}{\gamma}x + \frac{\alpha(\alpha+1)}{\gamma(\gamma+1)2!}x^2 + \cdots. \tag{1.42}$$

This result can be obtained from the case $\psi(x) = (1 - x)^\alpha$ by changing the variable $x \to x/\alpha$ and taking a limit $\alpha \to \infty$.

Chapter 2
Padé Approximation for P_{VI}

Abstract We will consider the Padé approximation problem associated with the function $\psi(x) = (1-x)^\alpha (1 - \frac{x}{t})^\beta$, and derive a pair of linear differential equations from it. Since the compatibility condition of the pair is the sixth Painlevé equation P_{VI}, we obtain special solutions of the P_{VI} equation from the Padé problem.

2.1 Derivation of the Differential Equation

In this chapter, we consider the following Padé approximation problem.

Find polynomials $P(x)$, $Q(x)$ of degree m, n such that

$$\psi(x) = \frac{P(x)}{Q(x)} + O(x^{m+n+1}), \qquad (2.1)$$

where

$$\psi(x) = (1-x)^\alpha \left(1 - \frac{x}{t}\right)^\beta. \qquad (2.2)$$

We will see that this problem is related to the sixth Painlevé equation P_{VI}. First, in this section, we show the following:

Theorem 2.1 *[63] The differential equation satisfied by $P(x)$ and $\psi(x)Q(x)$ is given explicitly as*

$$y'' + a(x)y' + b(x)y = 0, \qquad (2.3)$$

$$a(x) = -\left\{ \frac{m+n}{x} + \frac{\alpha-1}{x-1} + \frac{\beta-1}{x-t} + \frac{1}{x-q} \right\}, \qquad (2.4)$$

$$b(x) = \frac{1}{x(x-1)} \left\{ m(\alpha+\beta+n) + \frac{q(q-1)p}{x-q} - \frac{t(t-1)H}{x-t} \right\}, \qquad (2.5)$$

where q, p, H are certain parameters.

H. Nagao and Y. Yamada, *Padé Methods for Painlevé Equations*,
SpringerBriefs in Mathematical Physics,
https://doi.org/10.1007/978-981-16-2998-3_2
9

Later, the parameter H will be determined as a function of q, p, see (2.20). The parameters q, p are called an *accessory parameters* since they cannot be determined from the prescribed singularity data (at $x = 0, 1, t, \infty$). Such parameters will play the role of unknown variables of the Painlevé equation.

Proof As before, the equation can obtained from the Wronskian

$$\begin{vmatrix} y & y' & y'' \\ \mathbf{u} & \mathbf{u}' & \mathbf{u}'' \end{vmatrix} = |\mathbf{u}, \mathbf{u}'| y'' - |\mathbf{u}, \mathbf{u}''| y' + |\mathbf{u}', \mathbf{u}''| y = 0, \quad \mathbf{u} = \begin{bmatrix} P \\ \psi Q \end{bmatrix}. \tag{2.6}$$

Computation of $a(x)$. Noting that $\frac{\psi'}{\psi} = \frac{\alpha}{x-1} + \frac{\beta}{x-t}$, we have

$$|\mathbf{u}, \mathbf{u}'| = \psi \begin{vmatrix} P & P' \\ Q & \frac{\psi'}{\psi} Q + Q' \end{vmatrix} = \psi \begin{vmatrix} x^{[m]} & x^{[m-1]} \\ x^{[n]} & \dfrac{x^{[n+1]}}{(x-1)(x-t)} \end{vmatrix} = \psi \frac{x^{[m+n+1]}}{(x-1)(x-t)}. \tag{2.7}$$

The notation $x^{[k]}$ stands for a polynomial in x of degree at most k. Since this determinant is divisible by x^{m+n}, we have

$$|\mathbf{u}, \mathbf{u}'| = \psi x^{m+n} \frac{C(x-q)}{(x-1)(x-t)}, \tag{2.8}$$

where C, q are constants with respect to x. Accordingly, we have

$$-a(x) = \frac{|\mathbf{u}, \mathbf{u}''|}{|\mathbf{u}, \mathbf{u}'|} = \{\log(|\mathbf{u}, \mathbf{u}'|)\}' = \frac{m+n}{x} + \frac{\alpha-1}{x-1} + \frac{\beta-1}{x-t} + \frac{1}{x-q}. \tag{2.9}$$

Computation of $b(x)$. Since $\psi'' = \frac{x^{[2]}}{(x-1)^2(x-t)^2}\psi$, we have

$$|\mathbf{u}', \mathbf{u}''| = \psi x^{m+n-1} \frac{x^{[2]}}{(x-1)^2(x-t)^2}, \tag{2.10}$$

and hence

$$b(x) = \frac{|\mathbf{u}', \mathbf{u}''|}{|\mathbf{u}, \mathbf{u}'|} = \frac{C_0 + C_1 x + C_2 x^2}{x(x-1)(x-t)(x-q)}. \tag{2.11}$$

From an existence of a polynomial solution P of degree m, the constant C_2 is determined as $C_2 = m(\alpha + \beta + n)$. It is convenient to parametrize the expression (2.11) by the residues p and $-H$ at $x = q$ and $x = t$ respectively, then $b(x)$ takes the desired form (2.5). The parameter H will be determined as a function of p, q (see Proposition 2.1). $\qquad\square$

The Riemann scheme of (2.3) is

$x = 0$	1	t	q	∞
0	0	0	0	$-m$
$m + n + 1$	α	β	2	$-\alpha - \beta - n$

$$(2.12)$$

This can be seen directly from the singularity structure of the solutions $P(x)$, $\psi(x)Q(x)$. The singularity at $x = q$ is called an *apparent singularity* since the solutions are regular there. Note that the exponents at $x = q$ are $(0,2)$ while the exponents at ordinary regular points are $(0,1)$. In this example such an apparent singular point is needed to fulfill the Fuchs relation.

Non-logarithmic condition

We make a brief remark on the case where the exponents have an integer difference. A 2nd-order equation (1.1) with regular singular point at $x = q$ with exponents ρ_1, ρ_2 is expected to have two linearly independent power series solutions

$$y_i = (x - q)^{\rho_i} \sum_{n=0}^{\infty} c_{i,n}(x - q)^n \quad (i = 1, 2) \tag{2.13}$$

around $x = q$. This is true for generic exponents ρ_1, ρ_2 where the coefficients $\{c_{1,n}\}$ and $\{c_{2,n}\}$ can be determined through the linear recursion relations separately. However, we need some additional considerations if $\rho_1 - \rho_2 \in \mathbb{Z}$.

We assume $\rho_2 = \rho_1 + k$ ($k \in \mathbb{Z}_{\geq 0}$). There is no problem for the solution y_2. However, for the solution y_1, there may be a trouble since the recursion relation for $\{c_{1,n}\}$ will get mixed up with that for $\{c_{2,n}\}$. In this case, the recursion relation for $\{c_{1,n}\}$ has the form

$$n(n - k)c_{1,n} = R_n \quad (n > 0), \tag{2.14}$$

where R_n is a certain linear function of $c_{1,0}, c_{1,1}, \ldots, c_{1,n-1}$. We have two cases.

(Case i) $R_k = 0$. One can solve for $\{c_{1,n}\}$ with two free parameters $c_{1,0}$ and $c_{1,k}$. In this case, we still have two linear independent solutions in the power series form.

(Case ii) $R_k \neq 0$. There is no solution for $c_{1,k}$. This means that there is no power series solution y_1 in (2.13). Hence, the ansatz for y_1 in (2.13) is not available and we replace it by including a logarithmic term[1]

$$y_1 = (x - \alpha)^{\rho_1} \sum_{n=0}^{\infty} c_{1,n}(x - \alpha)^n + \text{Const.} y_2 \log(x - \alpha). \tag{2.15}$$

(Case i) occurs only when the coefficients of the differential equations satisfy a special relation (the *non-logarithmic condition*).

For instance, if $x = q$ is the apparent singularity with exponents $(\rho_1, \rho_2) = (0, 2)$, we should have two holomorphic solutions of the form

---

[1] See [67] Sect. 10.32. A convenient way to obtain such a solution is known as the Frobenius method: consider a regularized "solution" for $\rho = \rho_2 + \epsilon$ and expanding (or taking the derivative) w.r.t ϵ.

$$y_1 = 1 + O(z), \quad y_2 = z^2(1 + O(z)), \tag{2.16}$$

where $z = x - q$. This means that the coefficient c_2 of the series solution

$$y_1 = 1 + c_1 z + c_2 z^2 + \cdots, \tag{2.17}$$

should be a free parameter. As an example, let us put the expansion coefficients of $a(x), b(x)$ around $z = 0$ as

$$a(x) = z^{-1}(-1 + a_1 z + \cdots), \quad b(x) = z^{-2}(0 + b_1 z + b_2 z^2 + \cdots). \tag{2.18}$$

Putting these into (2.3), we see $c_1 = b_1$, and c_2 is a free parameter if and only if

$$a_1 b_1 + b_2 + b_1^2 = 0. \tag{2.19}$$

This is the non-logarithmic condition in the case of exponents $(\rho_1, \rho_2) = (0, 2)$. For the differential equation (2.3) arising from the Padé problem, this condition should be satisfied since the solutions $P(x), \psi(x)Q(x)$ are non-logarithmic at $x = q$.

Proposition 2.1 *From the non-logarithmic condition at $x = q$, the parameter H in (2.5) is determined as*

$$H = \frac{q(q-1)(q-t)}{t(t-1)}\left\{ p^2 - \left(\frac{m+n+1}{q} + \frac{\alpha}{q-1} + \frac{\beta-1}{q-t} \right)p + \frac{m(\alpha+\beta+n)}{q(q-1)} \right\}. \tag{2.20}$$

Proof From (2.5), we have expansions (2.18) where

$$a_1 = -\frac{m+n}{q} - \frac{\alpha-1}{q-1} - \frac{\beta-1}{q-t}, \quad b_1 = p, \tag{2.21}$$

$$b_2 = -\frac{t(t-1)H}{q(q-1)(q-t)} + \frac{m(\alpha+\beta+n)}{q(q-1)} - \frac{(2q-1)p}{q(q-1)},$$

hence from (2.19) we obtain (2.20). □

2.2 Deformation Equation

In the same situation as the previous section, we compute another linear equation between $y, y', y_t = \frac{\partial y}{\partial t}$ satisfied by $P(x)$ and $\psi(x)Q(x)$. Here, to fix the normalization of the polynomials $P(x), Q(x)$, we put $P(0) = 1$ (hence $Q(0) = 1$).

Proposition 2.2 *We have*

$$\frac{t(t-1)(x-q)}{t-q}y_t + x(x-1)y' - (q-1)pxy = 0. \tag{2.22}$$

Proof The equation is given by the following determinant:

$$\begin{vmatrix} y & y' & y_t \\ \mathbf{u} & \mathbf{u}' & \mathbf{u}_t \end{vmatrix} = |\mathbf{u}, \mathbf{u}'|y_t - |\mathbf{u}, \mathbf{u}_t|y' + |\mathbf{u}', \mathbf{u}_t|y = 0. \tag{2.23}$$

The determinant $|\mathbf{u}, \mathbf{u}'|$ is already computed in (2.8), and we will compute the remaining two.

Since $\frac{\psi_t}{\psi} = \frac{\beta x}{t(t-x)}$, we have

$$|\mathbf{u}, \mathbf{u}_t| = \psi \begin{vmatrix} P & P_t \\ Q & \frac{\psi_t}{\psi}Q + Q_t \end{vmatrix} = \psi \begin{vmatrix} x^{[m]} & x^{[m]} \\ x^{[n]} & \dfrac{x^{[n+1]}}{x-t} \end{vmatrix} = \psi \frac{x^{[m+n+1]}}{x-t}. \tag{2.24}$$

Since this is divisible by x^{m+n+1}, we have

$$|\mathbf{u}, \mathbf{u}_t| = c_1 \psi \frac{x^{m+n+1}}{x-t}, \tag{2.25}$$

where c_1 is a constant.

Next, for the determinant $|\mathbf{u}', \mathbf{u}_t|$, we have

$$|\mathbf{u}', \mathbf{u}_t| = \psi \begin{vmatrix} P' & P_t \\ \dfrac{\psi'}{\psi}Q + Q' & \dfrac{\psi_t}{\psi}Q + Q_t \end{vmatrix} \tag{2.26}$$

$$= \psi \begin{vmatrix} x^{[m-1]} & x^{[m]} \\ \dfrac{x^{[n+1]}}{(x-1)(x-t)} & \dfrac{x^{[n+1]}}{x-t} \end{vmatrix} = \psi \frac{x^{[m+n+1]}}{(x-1)(x-t)}.$$

Due to the normalization $P(0) = 1$, $|\mathbf{u}', \mathbf{u}_t|$ is divisible by x^{m+n+1}, and we have

$$|\mathbf{u}', \mathbf{u}_t| = c_2 \psi \frac{x^{m+n+1}}{(x-1)(x-t)}, \tag{2.27}$$

where c_2 is a constant. Thus we have the following relation:

$$C(x-q)y_t + c_1 x(x-1)y' + c_2 xy = 0. \tag{2.28}$$

The ratios of the constants C, c_1, c_2 are determined by the following conditions.

(i) From (2.3), we see that $y' = py$ for $x = q$. (ii) From the regularity of $P(x)$, $Q(x)$ at $x = t$, we have

$$\lim_{x \to t} \frac{|\mathbf{u}, \mathbf{u}_t|}{|\mathbf{u}, \mathbf{u}'|} = \lim_{x \to t} \frac{\psi_t}{\psi_x} = 1. \tag{2.29}$$

As a result, we have the desired Eq. (2.22).   □

2.3   Explicit Solutions by Schur Functions

From the solution of the Padé problem (2.1), the parameters q (and p also) in Eqs. (2.3), (2.22) can be determined as a function of t. By construction, these equations are compatible for such q (and p). Since the compatibility condition is equivalent to the P_{VI} equation (see Sect. 2.6), we see that the function $q(t)$ determined by Padé problem gives a special solution for P_{VI}. Here, we will carry out the computation of $q(t)$.

An explicit formula of the polynomials $P(x)$, $Q(x)$ is given by determinants (Schur functions). This formula is useful to obtain special solutions of various Painlevé type equations.

We consider the Padé problem (2.1) for

$$\psi(x) = \sum_{k=0}^{\infty} p_k x^k, \tag{2.30}$$

($p_0 = 1$, $p_i = 0, i < 0$). We define the *Schur function* s_λ corresponding to the sequence of integers $\lambda = (\lambda_1, \lambda_2, \cdots)$ by the *Jacobi–Trudi formula*

$$s_{(\lambda_1, \ldots, \lambda_l)} = \det(p_{\lambda_i - i + j})^l_{i,j=1}, \tag{2.31}$$

e.g.

$$s_{(3,3,1)} = \begin{array}{|c|c|c|} \hline & & \\ \hline & & \\ \hline & & \\ \hline \end{array} = \begin{vmatrix} p_3 & p_4 & p_5 \\ p_2 & p_3 & p_4 \\ 0 & 1 & p_1 \end{vmatrix}. \tag{2.32}$$

Theorem 2.2 *The polynomials $P(x)$ and $Q(x)$ are given by*

$$P(x) = \sum_{i=0}^{m} s_{(m^n, i)} x^i, \quad Q(x) = \sum_{i=0}^{n} s_{((m+1)^i, m^{n-i})} (-x)^i, \tag{2.33}$$

where a sequence of integers $(\underbrace{m, m, \ldots, m}_{n})$ is abbreviated as m^n.

Example

In the case of $(m, n) = (3, 2)$, we have

$$P(x) = \boxed{} + \boxed{}\, x + \boxed{}\, x^2 + \boxed{}\, x^3, \tag{2.34}$$

$$Q(x) = \boxed{} - \boxed{}\, x + \boxed{}\, x^2. \tag{2.35}$$

The normalization of $P(x)$, $Q(x)$ in (2.33) is such that $P(0) = Q(0) = \tau_{m,n}$, where

$$\tau_{m,n} = s_{m^n} \tag{2.36}$$

is the Schur function corresponding to the rectangular Young diagram m^n.

Proof of Theorem 2.2. Let us check that the polynomials $P(x)$, $Q(x)$ given by (2.33) satisfy the Padé approximation condition

$$\psi(x)Q(x) - P(x) = O(x^{m+n+1}). \tag{2.37}$$

The polynomial $Q(x)$ can be written as

$$Q(x) = \begin{vmatrix} p_m & p_{m+1} & \cdots & \cdots & p_{m+n} \\ \cdots & p_m & p_{m+1} & \cdots & \cdots \\ & & \ddots & \ddots & \\ \cdots & \cdots & \cdots & p_m & p_{m+1} \\ x^n & \cdots & \cdots & x & 1 \end{vmatrix}, \tag{2.38}$$

Hence we have

$$\psi(x)Q(x) = \sum_{i=1}^{\infty} \begin{vmatrix} p_m & p_{m+1} & \cdots & \cdots & p_{m+n} \\ \cdots & p_m & p_{m+1} & \cdots & \cdots \\ & & \ddots & \ddots & \\ \cdots & \cdots & \cdots & p_m & p_{m+1} \\ p_{i-n} & \cdots & \cdots & p_{i-1} & p_i \end{vmatrix} x^i = \sum_{i=0}^{\infty} s_{(m^n, i)} x^i \tag{2.39}$$

$$= \sum_{i=0}^{m} s_{(m^n, i)} x^i + \sum_{i=m+1}^{\infty} s_{(m^n, i)} x^i = P_m(x) + O(x^{m+n+1}),$$

as desired. $\square$

Applying the formulas (2.33), we derive an explicit form of a special solution $q(t)$ of P_{VI}.

Theorem 2.3 *The following function gives a special solution of P_{VI} for $m, n \in \mathbb{Z}_{\geq 0}$:*

$$q(t) = \frac{t(m+n+1)}{m-n-\alpha-\beta} \frac{\tau_{m,n}\tau_{m+1,n+1}}{\tau_{m+1,n}\tau_{m,n+1}}. \tag{2.40}$$

Proof To obtain the explicit form of the parameter q in (2.8), we put

$$|\mathbf{u}, \mathbf{u}'| = \psi(1-x)^{-1}(1-x/t)^{-1}x^{m+n}(A+Bx). \tag{2.41}$$

From (2.39) we have $\psi Q - P = vx^{m+n+1} + O(x^{m+n+2})$, where

$$v = s_{(m^n, m+n+1)} = (-1)^n \tau_{m+1,n+1}. \tag{2.42}$$

Then we have

$$|\mathbf{u}, \mathbf{u}'| = \begin{vmatrix} P & P' \\ \psi Q - P & (\psi Q - P)' \end{vmatrix} \tag{2.43}$$

$$= \begin{vmatrix} P(0) & P(1) \\ vx^{m+n+1} & (m+n+1)vx^{m+n} \end{vmatrix} + O(x^{m+n})$$

$$= (m+n+1)P(0)vx^{m+n} + O(x^{m+n+1}).$$

Hence

$$A = (-1)^n (m+n+1)\tau_{m,n}\tau_{m+1,n+1}. \tag{2.44}$$

On the other hand, for $x \to \infty$ we have

$$\psi^{-1}|\mathbf{u}, \mathbf{u}'| = \psi^{-1} \begin{vmatrix} P & P' \\ \psi Q & (\psi Q)' \end{vmatrix} \tag{2.45}$$

$$= \begin{vmatrix} \tau_{m,n+1} & m\tau_{m,n+1} \\ (-1)^n\tau_{m+1,n} & (-1)^n(\alpha+\beta+n)\tau_{m+1,n} \end{vmatrix} x^{m+n-1} + O(x^{m+n-2}),$$

and hence

$$B = (-1)^n \frac{(\alpha+\beta-m+n)}{t} \tau_{m+1,n}\tau_{m,n+1}. \tag{2.46}$$

Thus we get

$$q = -\frac{B}{A} = \frac{t(m+n+1)}{m-n-\alpha-\beta} \frac{\tau_{m,n}\tau_{m+1,n+1}}{\tau_{m+1,n}\tau_{m,n+1}}, \tag{2.47}$$

as desired. $\square$

Remark 2.1 The function $\psi(x)$ (2.2) has an expansion

$$\psi(x) = (1-x)^\alpha (1-\frac{x}{t})^\beta = \sum_{i,l\geq 0} \frac{(-\alpha)_i}{i!} \frac{(-\beta)_l}{l!} t^{-l} x^{i+l}, \qquad (2.48)$$

where $(\alpha)_k = \alpha(\alpha+1)\cdots(\alpha+k-1)$. Hence the coefficient p_k can be written as

$$p_k = \sum_{l=0}^{k} \frac{(-\alpha)_{k-l}}{(k-l)!} \frac{(-\beta)_l}{l!} t^{-l} = \frac{(-\alpha)_k}{k!} \sum_{l=0}^{k} \frac{(-k)_l(-\beta)_l}{(\alpha-k+1)_l l!} t^{-l} \qquad (2.49)$$

$$= \frac{(-\alpha)_k}{k!} {}_2F_1(-k,-\beta,\alpha-k+1;t^{-1}).$$

Thus we obtained the explicit form of the special solution (2.40) of P_{VI} given in terms of the determinants of Jacobi polynomials. Though such solutions are well known, the derivation [63] explained here is quite simple.

$\tau_{m,n}$ as the τ-function

The most fundamental object in the isomonodromic deformation is the τ-function [20–22]. It can be defined as $\frac{d}{dt}\log\tau = H(q(t),p(t),t)$ up to a normalization. More precisely, for a solution p,q for the P_{VI} Eq. (2.97) with $k = \frac{(b+c+d)^2-a^2}{4}$, the function σ defined by

$$\sigma = t(t-1)H(q,p) + k_1 t + k_2, \qquad (2.50)$$

where $k_1 = \frac{d(d+2b+2c)-a^2}{4}$, $k_2 = \frac{a^2-b^2+c^2-4bd-d^2}{8}$ satisfies the following equation called the σ-form [21, 45]:

$$\sigma'(t(t-1)\sigma'')^2 + (2\sigma'(t\sigma'-\sigma) - (\sigma')^2 + c_1c_2c_3c_4)^2 - \prod_{i=1}^{4}(\sigma'-c_i^2) = 0, \qquad (2.51)$$

where $' = \frac{d}{dt}$ and $(c_1,\cdots,c_4) = \frac{1}{2}(a+d,a-d,b+c,b-c)$.

For the special solutions obtained from the Padé method, the determinants $\tau_{m,n}$ give the τ-function. In fact, one can check that the following function

$$\sigma = t(t-1)(\log\tau_{m,n})' + (1-t)c_1^2 + \frac{1}{2}\sum_{1\leq i<j\leq 4} c_ic_j, \qquad (2.52)$$

satisfies (2.51) with $(c_1,\ldots,c_4) = (\frac{-\alpha+m-n}{2}, \frac{\alpha-m-n}{2}, \frac{\alpha+m+n}{2}, \frac{\alpha-m+n}{2} + \beta)$. For the derivation of similar and more general formulae, see [32] for example.

Remark 2.2 We have seen the connection between certain Padé approximations and isomonodromic equations. In fact, it has been known that there are close connections among the orthogonal polynomials, Padé approximations, and Painlevé equations (see [4, 28, 30, 31, 62] for example). Our main strategy is to formulate and study various differential/discrete isomonodromic equations starting suitable Padé problems. We will find this is a useful method since the Padé problem is easy to attack.

2.4 Extension to Garnier System

Soon after the work by R. Fuchs, an extension of the P_{VI} was obtained by Garnier [8] as the isomonodromic deformation of the Fuchsian equation on $\mathbb{P}^1$ with $N + 3$ regular singular points at $\{0, 1, t_1, \ldots, t_N, \infty\}$ (together with N apparent singularities $\{q_1, \ldots, q_N\}$). This system can be written as a multi-time Hamiltonian system with $2N$ unknown variables p_i, q_i and N time variables t_i ($i = 1, \ldots, N$), and called the *N-Garnier system* [8, 9] (see [17] for modern exposition). For convenience of the readers, we give a short introduction in Sect. 2.6. The Garnier system includes the P_{VI} as the $N = 1$ case.

To obtain the N-Garnier system, we consider the following Padé problem.

Find polynomials $P(x)$, $Q(x)$ of degree m, n such that

$$\psi(x) = \frac{P(x)}{Q(x)} + O(x^{m+n+1}), \tag{2.53}$$

where

$$\psi(x) = (1 - x)^{\kappa} \prod_{i=1}^{N} (1 - \frac{x}{t_i})^{\alpha_i}. \tag{2.54}$$

In this case, the Riemann scheme is given by

$x = 0$	1	t_1	$\cdots$	t_N	∞	λ_1	$\cdots$	λ_N	
0	0	0	$\cdots$	0	$-m$	0	$\cdots$	0	, (2.55)
$m+n+1$	κ	α_1	$\cdots$	α_N	$-n - \kappa - \sum_{i=1}^{N} \alpha_i$	2	$\cdots$	2	

where $\lambda_1, \ldots, \lambda_N$ are apparent singularities.

Coefficient $a(x)$ of the differential equation $L : y'' + a(x)y' + b(x)y = 0$ has the form

$$-a(x) = \frac{m+n}{x} + \frac{\kappa - 1}{x - 1} + \sum_{i=1}^{N} \frac{\alpha_i - 1}{x - t_i} + \sum_{i=1}^{N} \frac{1}{x - \lambda_i}. \tag{2.56}$$

Define variables μ_i and K_i by

$$\mu_i = \operatorname*{Res}_{x=\lambda_i} b(x), \quad K_i = -\operatorname*{Res}_{x=t_i} b(x). \tag{2.57}$$

By the non-logarithmic condition at $x = \lambda_i$, K_i is determined as a rational function of $\{\lambda_j, \mu_j\}$.

We also have the deformation equations of the form

$$\{\partial_{t_j} + G_j(x)\partial_x + F_j(x)\}y(x) = 0 \quad (j = 1, \ldots, N), \tag{2.58}$$

then, as the compatibility equations, we have (see Sect. 2.6)

$$\frac{\partial \lambda_i}{\partial t_j} = \frac{\partial K_j}{\partial \mu_i}, \quad \frac{\partial \mu_i}{\partial t_j} = -\frac{\partial K_j}{\partial \lambda_i} \quad (i, j = 1, 2, \ldots, N). \tag{2.59}$$

This *multi-time Hamiltonian system* is the N-Garnier system.

The expansion of the function $\psi(x)$ (2.54) is given by

$$\psi(x) = \sum_{l=0}^{\infty} \frac{(-\kappa)_l}{(1)_l} x^l \cdot \sum_{m_i \geq 0} x^{|m|} \prod_{i=1}^{N} \frac{(-\alpha_i)_{m_i}}{(1)_{m_i}} t_i^{-m_i}, \tag{2.60}$$

where $(a)_k = a(a+1)\cdots(a+k-1)$ and $|m| = \sum_{i=1}^{N} m_i$. Hence we have

$$p_k = \frac{(-\kappa)_k}{(1)_k} F_D(-k, -\alpha_1, \ldots, -\alpha_N, \kappa - k + 1; t_1^{-1}, \ldots, t_N^{-1}), \tag{2.61}$$

where F_D is the *Appell–Lauricella function*

$$F_D(\alpha, \beta_1, \ldots, \beta_N, \gamma; x_1, \ldots, x_N) = \sum_{m_i \geq 0} \frac{(\alpha)_{|m|}}{(\gamma)_{|m|}} \left\{ \prod_{i=1}^{N} \frac{(\beta_i)_{m_i}}{(1)_{m_i}} x_i^{m_i} \right\}. \tag{2.62}$$

In order to specify $\{\lambda_i\}$ as functions of $\{t_j\}$ (or to give special solutions of the Garnier systems), it is convenient to consider the polynomial $F(x) = \prod_{i=1}^{N}(x - \lambda_i)$. Since $F(x)$ appears as the non-trivial factor of the Wronskian $|\mathbf{u}, \mathbf{u}'|$, it can be explicitly determined by the following special values at $x = t_i$ ($i = 0, 1, \ldots, N$):

$$F(t_i) = \frac{a_i \prod_{j(\neq i)=0}^{N}(t_i - t_j)}{n - m + \sum_{i=0}^{N} a_i} \frac{(T_{\alpha_i} \tau_{m+1,n})(T_{\alpha_i}^{-1} \tau_{m,n+1})}{\tau_{m+1,n} \tau_{m,n+1}}, \tag{2.63}$$

where $t_0 = 1$, $\alpha_0 = \kappa$ and T_{α_i} means the shift $\alpha_i \to \alpha_i + 1$. We also have

$$F(0) = \frac{(m + n + 1) \prod_{j=0}^{N}(-t_j)}{n - m + \sum_{i=0}^{N} a_i} \frac{\tau_{m+1,n} \tau_{m,n+1}}{\tau_{m+1,n} \tau_{m,n+1}}. \tag{2.64}$$

The Eq. (2.63) can be derived from

$$P(t_i) = t_i^m T_{a_i} \tau_{m,n+1}, \quad Q(t_i) = (-t_i)^n T_{a_i}^{-1} \tau_{m+1,n}, \tag{2.65}$$

which follows from the formulae (2.69) in the next section.

The corresponding τ-functions for these special solutions are given by the determinants $\tau_{m,n}$ (see for example [59] where more general results are discussed). We will study the N-Garnier system further in Sect. 6.2.

2.5 More on the Schur Functions

We will discuss some further properties of Schur functions which play an important role in various areas of mathematical physics ([34] see also [60] for further generalization).

For any partition $\lambda = (\lambda_1, \lambda_2, \ldots, \lambda_\ell)$, the Schur functions s_λ can be viewed as minor determinants of the infinite matrix $X = (p_{j-i})_{i,j \in \mathbb{Z}}$:

$$X = \begin{bmatrix} \ddots & \ddots & \ddots & \cdots & \cdots & \\ & p_0 & p_1 & p_2 & \cdots & \\ & & p_0 & p_1 & p_2 & \cdots \\ & & & p_0 & p_1 & p_2 & \cdots \\ & & & & p_0 & p_1 & p_2 & \cdots \\ & O & & & & \ddots & \ddots & \ddots & \cdots & \cdots \end{bmatrix}, \tag{2.66}$$

where the row and column indices are chosen as $I = (1 - \ell, 2 - \ell, \ldots, -1, 0)$ and $J = \lambda - (0, 1, \ldots, \ell - 1)$ respectively. The index set J is conveniently represented by a figure called a *Maya diagram*.

Example

For $\lambda = (5, 3, 3, 2, 0)$, we have $I = (-4, -3, -2, -1, 0)$, $J = (5, 2, 1, -1, -4)$, hence the minor determinant gives the Schur function s_λ:

$$
\begin{array}{c|ccccc}
i\backslash j & -4 & -1 & 1 & 2 & 5 \\
\hline
-4 & p_0 & p_3 & p_5 & p_6 & p_9 \\
-3 & & p_2 & p_4 & p_5 & p_8 \\
-2 & & p_1 & p_3 & p_4 & p_7 \\
-1 & & p_0 & p_2 & p_3 & p_6 \\
0 & & & p_1 & p_2 & p_5 \\
\end{array}
=
\begin{vmatrix}
p_5 & p_6 & p_7 & p_8 \\
p_2 & p_3 & p_4 & p_5 \\
p_1 & p_2 & p_3 & p_4 \\
p_0 & p_1 & p_2 \\
\end{vmatrix}
= s_\lambda. \tag{2.67}
$$

In the first equality we used the fact that, for any $N \times N$ matrix A, a transformation $A \mapsto J A^T J$, $J = (\delta_{i+j,N})_{i,j=1}^{N}$ (i.e. the transposition along the anti-diagonal) preserves the determinant.

Since $s_\lambda = s_{(\lambda,0)}$, one can consider that the partition $\lambda = (\lambda_1, \lambda_2, \ldots, 0, 0, \cdots)$ is of infinite length. Then the index set $J = (5, 2, 1, -1, -4, -5, \cdots)$ and the corresponding Maya diagram is as follows:

We introduce operators which create/annihilate $\bullet$'s. For a parameter $x \in \mathbb{C}$, let $V^*(x)$, $V(x)$ be shift operators acting on a polynomial f in variables $\{p_i\}$ as

$$V^*(x)f = f\Big|_{p_i \to \sum_{j=0}^{i} x^{-j} p_{i-j}}, \qquad V(x)f = f\Big|_{p_i \to p_i - x^{-1} p_{i-1}}. \tag{2.68}$$

Proposition 2.3 *The polynomials $P(x)$ and $Q(x)$ in (2.33) can be expressed as*

$$P(x) = x^m V^*(x)\tau_{m,n+1}, \qquad Q(x) = (-x)^n V(x)\tau_{m+1,n}. \tag{2.69}$$

Proof (i) Put $\hat{p}_i = \sum_{j=0}^{i} x^{-j} p_{i-j}$ and consider the determinant

$$V^*(x)\tau_{m,n+1} = \begin{vmatrix} \hat{p}_m & \hat{p}_{m+1} & \cdots & \hat{p}_{m+n} \\ \hat{p}_{m-1} & \hat{p}_m & \cdots & \hat{p}_{m+n-1} \\ & \ddots & \ddots & \\ \hat{p}_{m-n} & \cdots & \hat{p}_{m-1} & \hat{p}_m \end{vmatrix}. \tag{2.70}$$

By the relation $\hat{p}_i - x^{-1}\hat{p}_{i-1} = p_i$, the $\hat{p}_i$'s in the determinant can be restored to p_i except for the last row. Hence we have

$$V^*(x)\tau_{m,n+1} = \sum_{i=0}^{m} s_{(m^n, m-j)} x^{-j} = x^{-m} P(x). \tag{2.71}$$

(ii) Put $\hat{p}_i = p_i - x^{-1} p_{i-1}$ and consider the determinant

$$V(x)\tau_{m+1,n} = \begin{vmatrix} \hat{p}_{m+1} & \hat{p}_{m+2} & \cdots & \hat{p}_{m+n} \\ \hat{p}_m & \hat{p}_{m+1} & \cdots & \hat{p}_{m+n-1} \\ & \ddots & \ddots & \\ \hat{p}_{m-n} & \cdots & \hat{p}_m & \hat{p}_{m+1} \end{vmatrix}. \tag{2.72}$$

Expanding $\hat{p}_i$ into (a) p_i or (b) $-x^{-1}p_{i-1}$ in each row, non-zero results arise only for the choices $(a)\ldots(a)(b)\ldots(b)$ from top to bottom, since the term vanishes for the choice $\ldots(b)(a)\ldots$. Hence we have

$$V(x)\tau_{m+1,n} = \sum_{i=0}^{n} s_{((m+1)^i,m^{n-i}}(-x)^{-(n-i))} = (-x)^{-n}Q(x), \qquad (2.73)$$

as desired.   □

Proposition 2.4 *For any partition* $\lambda = (\lambda_1 \geq \lambda_2 \geq \ldots)$, *we have*

$$V^*(x)s_\lambda = \psi(x)\sum_{i=1}^{\infty}(-1)^{i-1}s_{(\lambda_1+1,\lambda_2+1,\ldots,\lambda_{i-1}+1,\check{\lambda}_i,\lambda_{i+1},\lambda_{i+2},\ldots)}x^{i-1-\lambda_i}, \quad (2.74)$$

$$V(x)s_\lambda = \psi(x)^{-1}\sum_{k\in\mathbb{Z}}s_{(k,\lambda)}x^k, \qquad (2.75)$$

where $(\ldots, i, \check{j}, k, \ldots) = (\ldots, i, j, \ldots)$ *and* $(k, \lambda) = (k, \lambda_1, \lambda_2, \ldots)$.

Proof First consider (2.74). By the column operations using $\hat{p}_i - x^{-1}\hat{p}_{i-1} = p_i$, we have

$$V^*(x)s_\lambda =
\begin{vmatrix}
\hat{p}_{\lambda_1} & \hat{p}_{\lambda_1+1} & \cdots & \hat{p}_{\lambda_1+\ell-1} \\
\hat{p}_{\lambda_2-1} & \hat{p}_{\lambda_2} & \cdots & \hat{p}_{\lambda_2+\ell-2} \\
\vdots & & \ddots & \\
& & & \hat{p}_{\lambda_\ell}
\end{vmatrix}
=
\begin{vmatrix}
\hat{p}_{\lambda_1} & p_{\lambda_1+1} & \cdots & p_{\lambda_1+\ell-1} \\
\hat{p}_{\lambda_2-1} & p_{\lambda_2} & \cdots & p_{\lambda_2+\ell-2} \\
\vdots & & \ddots & \\
& & & p_{\lambda_\ell}
\end{vmatrix} \quad (2.76)$$

$$=
\begin{vmatrix}
\hat{p}_{\lambda_1} + \sum_{i=1}^{\ell-1}p_{\lambda_1+i}x^i & p_{\lambda_1+1} & \cdots & p_{\lambda_1+\ell-1} \\
\hat{p}_{\lambda_2-1} + \sum_{i=1}^{\ell-1}p_{\lambda_2+i-1}x^{i-1} & p_{\lambda_2} & \cdots & p_{\lambda_2+\ell-2} \\
\vdots & & \ddots & \\
& & & p_{\lambda_\ell}
\end{vmatrix}.$$

We consider that the partition $\lambda = (\lambda_1, \lambda_2, \ldots, 0, 0, 0)$ is of large length ℓ, and take the limit $\ell \to \infty$, then we have

$$V^*(x)s_\lambda \to \psi(x)
\begin{vmatrix}
x^{-\lambda_1} & p_{\lambda_1+1} & p_{\lambda_1+2} & \cdots \\
x^{-\lambda_2+1} & p_{\lambda_2} & p_{\lambda_2+1} & \cdots \\
x^{-\lambda_3+2} & p_{\lambda_3-1} & p_{\lambda_3} & \cdots \\
\vdots & & & \ddots
\end{vmatrix} \quad (2.77)$$

$$= \psi(x)\sum_{i=1}^{\infty}(-1)^i x^{i-1-\lambda_i}s_{\lambda_1+1,\lambda_2+1,\ldots,\lambda_{i-1}+1,\lambda_{i+1},\lambda_{i+2},\ldots)},$$

hence the Eq. (2.74) is proved. For (2.75), we have

$$\sum_{k\in\mathbb{Z}} s_{(k,\lambda)}x^k = \sum_{k\in\mathbb{Z}} \begin{vmatrix} p_k & p_{k+1} & \cdots \\ p_{\lambda_1-1} & p_{\lambda_1} & \cdots \\ \vdots & & \ddots \\ & & & p_{\lambda_\ell} \end{vmatrix} x^k = \psi(x) \begin{vmatrix} 1 & x^{-1} & \cdots \\ p_{\lambda_1-1} & p_{\lambda_1} & \cdots \\ \vdots & & \ddots \\ & & & p_{\lambda_\ell} \end{vmatrix}$$

$$= \psi(x)\{s_\lambda - s_{(\lambda_1-1,\lambda_2,\dots)}x^{-1} + s_{(\lambda_1-1,\lambda_2-1,\lambda_3,\dots)}x^{-2} - \cdots\} \tag{2.78}$$
$$= \psi(x)V(x)s_\lambda,$$

hence we obtain (2.75). $\square$

Since Schur functions are minor determinants, there are various *bilinear relations* among them.

Proposition 2.5 *We have*

$$s_{(k,\lambda)}.s_{(l-1,m,\lambda)} + s_{(l,\lambda)}.s_{(m-1,k,\lambda)} + s_{(m,\lambda)}.s_{(k-1,l,\lambda)} = 0, \tag{2.79}$$
$$V(a)s_\lambda.s_{(l-1,m,\lambda)} - as_{(l,\lambda)}.V(a)s_{(m,\lambda)} + as_{(m,\lambda)}.V(a)s_{(l,\lambda)} = 0, \tag{2.80}$$
$$bV(a)s_\lambda.V(b)s_{(m,\lambda)} - aV(b)s_\lambda.V(a)s_{(m,\lambda)} \tag{2.81}$$
$$+ (a-b)s_{(m,\lambda)}.V(a)V(b)s_\lambda = 0,$$
$$(b-c)V(a)s_\lambda.V(b)V(c)s_\lambda + (c-a)V(b)s_\lambda.V(c)V(a)s_\lambda \tag{2.82}$$
$$+ (a-b)V(c)s_\lambda.V(a)V(b)s_\lambda = 0.$$

Proof Though these also can be proved using fermion operators, we will give more elementary derivations. The Eq. (2.79) follows from the identity (*Plücker relation*)

$$d_{12}d_{34} + d_{23}d_{14} + d_{31}d_{24} = 0, \tag{2.83}$$

for determinants $d_{ij} = |\mathbf{a}_i, \mathbf{a}_j, \mathbf{x}_1, \dots, \mathbf{x}_n|$, $\mathbf{a}_i, \mathbf{x}_i \in \mathbb{C}^n$. The remaining relations (2.80)–(2.82) can be derived from (2.79) using $s_{(m-1,k,\lambda)} = -s_{(k-1,m,\lambda)}$, $\sum_{k\in\mathbb{Z}} a^k s_{(k,\lambda)} = \psi(a)V(a)s_\lambda$, $\sum_{l\in\mathbb{Z}} b^l V(a)s_{(l,\lambda)} = (1 - \frac{b}{a})\psi(b)V(a)V(b)s_\lambda$, etc. $\square$

Proposition 2.6 *In particular, for $\tau_{m,n} = s_{(m^n)}$, we have*

$$\tau_{m-1,n}.V(a)\tau_{m+1,n} + \tau_{m,n+1}.V(a)\tau_{m,n-1} - \tau_{m,n}.V(a)\tau_{m,n} = 0, \tag{2.84}$$
$$\tau_{m-1,n}.V(a)\tau_{m,n+1} - \tau_{m,n+1}.V(a)\tau_{m-1,n} + a^{-1}\tau_{m-1,n+1}.V(a)\tau_{m,n} = 0, \tag{2.85}$$
$$bV(a)\tau_{m,n}.V(b)\tau_{m,n+1} - aV(b)\tau_{m,n}.V(a)\tau_{m,n+1} \tag{2.86}$$
$$+ (a-b)\tau_{m,n+1}.V(a)V(b)\tau_{m,n} = 0.$$

Proof Put $(k,l,m) \to (m+1, l, -n+1)$ and $\lambda \to ((m-1)^{n-i-1}, m^i)$ in (2.79), we obtain the coefficients a^{-i} of (2.84). Putting $l \to -n$, $\lambda \to (m^n)$ in (2.80) we have (2.85). Putting $\lambda \to (m^n)$ in (2.81) we have (2.86). $\square$

Date's direct method

Here, we explain Date's direct method in soliton theory [5], which is based on a similar strategy to our method. Though Date's original work deals with certain nonlinear differential equations, we will discuss in a discrete setting to show the algebraic structure more transparently.

Let $\Psi_n(z)$ be a monic polynomial of degree n subjected to the linear relations

$$\Psi_n(p_i) = c_i \Psi_n(q_i) \quad (i = 1, \ldots, n). \tag{2.87}$$

For generic parameters p_i, q_i, c_i, the solution is unique. Let V_z be a shift operator acting on variables c_i as

$$V_z : c_i \mapsto \frac{z - q_i}{z - p_i} c_i. \tag{2.88}$$

Consider a polynomial in z of degree at most n defined by

$$\tilde{\Psi}_n(z) := (z - a) V_a \Psi_n(z) - (z - b) V_b \Psi_n(z) + (a - b) \frac{V_b(\Psi_n(a))}{\Psi_n(a)} \Psi_n(z). \tag{2.89}$$

It is easy to show that $\tilde{\Psi}_n(z)$ satisfies the same condition (2.87) as $\Psi_n(z)$, and hence $\tilde{\Psi}_n(z) = C \Psi_n(z)$ with C independent of z. Since $\tilde{\Psi}_n(z)$ vanishes at $z = a$, the constant C is zero, namely $\tilde{\Psi}_n(z) = 0$ identically.

Since $\tilde{\Psi}(b) = 0$ also, we have

$$\frac{V_b(\Psi_n(a))}{\Psi_n(a)} = \frac{V_a(\Psi_n(b))}{\Psi_n(b)}. \tag{2.90}$$

From this, we see that there exists a polynomial τ_n in $\{c_i\}$ (independent of z) such that

$$\Psi_n(z) = \frac{\prod_{i=1}^n (z - p_i) V_z(\tau_n)}{\tau_n}. \tag{2.91}$$

Then, from (2.89) and (2.91), we have the *Hirota–Miwa equation* [15]

$$(b - c) V_a(\tau_n) V_b V_c(\tau_n) + (abc \text{ cyclic}) = 0. \tag{2.92}$$

Explicitly, the τ-functions τ_n can be written as

$$\begin{aligned}
\tau_0 &= 1, \\
\tau_1 &= 1 + \eta_1, \\
\tau_2 &= 1 + \eta_1 + \eta_2 + a_{1,2} \eta_1 \eta_2, \\
\tau_3 &= 1 + \sum_{i=1}^3 \eta_i + \sum_{i<j} a_{i,j} \eta_i \eta_j + a_{1,2} a_{1,3} a_{2,3} \eta_1 \eta_2 \eta_3,
\end{aligned} \tag{2.93}$$

and determined recursively by

$$\tau_n = \tau_{n-1} + \eta_n V_{p_n}^{-1} V_{q_n}(\tau_{n-1}) = \tau_{n-1} + \eta_n (\tau_{n-1}|_{\eta_i \mapsto a_{i,n}\eta_i}), \qquad (2.94)$$

where

$$\eta_i = -c_i \prod_{j(\neq i)=1}^{n} \frac{p_i - p_j}{q_i - p_j}, \quad a_{i,j} = \frac{(p_i - p_j)(q_j - q_i)}{(p_i - q_j)(p_j - q_i)}. \qquad (2.95)$$

These are well-known formulas of the soliton solutions for various soliton equations.

This example shows that a suitable linear problem may give interesting solutions for nonlinear equations. Such a simple method to derive soliton solutions is known as Date's direct method, which can be viewed as a shortcut of the inverse scattering method, or the degenerate limit of Krichever's construction of the theta function solutions from the Baker–Akhiezer function.

Similarly, in our approach, the Padé approximations give suitable linear problems to study the isomonodromic equations.

2.6   Appendix

In this appendix, we recall some basic facts on Painlevé and Garnier equations.

P_{VI} equation

Painlevé equations are the very special nonlinear differential equations discovered by Painlevé and Gambier around 1900. Classically, they are classified into six equations P_J (J = I, ..., VI). The most generic case among them is the *sixth Painlevé equation* P_{VI} and others can be considered as its degenerations. For the general backgrounds of these equations we refer the reader to the textbook [17] and references therein. Here, we will give a very brief introduction to the equation P_{VI}.

The original form of P_{VI} takes the following form:

$$q'' = \frac{1}{2}\left(\frac{1}{q} + \frac{1}{q-1} + \frac{1}{q-t}\right)q'^2 - \left(\frac{1}{t} + \frac{1}{t-1} + \frac{1}{q-t}\right)q'$$
$$+ \frac{q(q-1)(q-t)}{t^2(t-1)^2}\left(\alpha + \beta\frac{1}{q^2} + \gamma\frac{t-1}{(q-1)^2} + \delta\frac{t(t-1)}{(q-t)^2}\right). \qquad (2.96)$$

Here $q = q(t)$ is the unknown function, $' = \frac{d}{dt}$, and $\alpha, \beta, \gamma, \delta$ are constant parameters. This equation can be written in a *Hamiltonian form*

$$q' = \frac{\partial H}{\partial p}, \quad p' = -\frac{\partial H}{\partial q}, \qquad (2.97)$$

with the Hamiltonian

$$H = \frac{q(q-1)(q-t)}{t(t-1)}\left\{p^2 + (\frac{b}{q} + \frac{c}{q-1} + \frac{d}{q-t})p + \frac{k}{q(q-1)}\right\}, \quad (2.98)$$

where $\alpha = \frac{(b+c+d)^2-4k}{2}$, $\beta = -\frac{b^2}{2}$, $\gamma = \frac{c^2}{2}$, $\delta = -\frac{d(d-2)}{2}$. The Hamiltonian form is fundamental for various studies of the P_{VI} equation. Though $H(p,q)$ (2.98) looks like a rational function of (p,q), it is actually a polynomial.

Eliminating the variable p from Eq. (2.97), one can recover (2.96). To carry out this easily, we recommend the following procedure. First, consider the general case where $H = H(p,q,t) = A(q,t)p^2 + B(q,t)p + C(q,t)$ to obtain

$$q'' = \frac{1}{2A}\frac{\partial A}{\partial q}q'^2 + \frac{1}{A}\frac{\partial A}{\partial t}q' + A\left\{\frac{\partial}{\partial t}(\frac{B}{A}) + \frac{\partial}{\partial q}(\frac{B^2}{2A} - 2C)\right\}. \quad (2.99)$$

Then, specializing A, B, C as in (2.98), derive the expansion of $\frac{B^2}{2A} - 2C$ as

$$\frac{A}{2}(\frac{B}{A})^2 - 2C = \frac{c_1}{q} + \frac{c_2}{q-1} + \frac{c_3}{q-t} + c_4 q + c_0, \quad (2.100)$$

where $c_1 = \frac{b^2}{2(t-1)}$, $c_2 = -\frac{c^2}{2t}$, $c_3 = \frac{d^2}{2}$, $c_4 = \frac{(b+c+d)^2-4k}{2t(t-1)}$. The key to this computation is that we do not need to know the coefficient c_0, which is a little complicated.

One of the most important aspects of the P_{VI} is its characterization as the isomonodromic deformation due to R. Fuchs [7]. Namely, P_{VI} describes the isomonodromic deformation of the Fuchsian equation on $\mathbb{P}^1$ with four regular singular points at $\{0, 1, t, \infty\}$.

Lax pair

In previous sections, from the Padé problem for $\psi(x)$ in (2.2), we derived two differential equations (2.3) and (2.22) of the form

$$Ly = 0: \quad L = \partial_x^2 + a(x)\partial_x + b(x), \quad (2.101)$$
$$By = 0: \quad B = \partial_t + G(x)\partial_x + F(x).$$

In relation to the Padé problem, the Eq. (2.101) can be considered in the following two different contexts:

- (Situation *on-Padé*) $m, n \in \mathbb{Z}_{\geq 0}$ and the parameters q, p are some fixed functions of α, β, m, n and t, specified by the Padé problem. In this situation, the equations are automatically compatible by construction.

> • (Situation *off-Padé*) $m, n \in \mathbb{C}$ and the parameters q, p are variables or unknown functions to be solved from the compatibility condition.
> One can obtain various results by switching these two situations suitably.

In this section, we study the Eq. (2.101) in an off-Padé situation. Then we have the following:

Theorem 2.4 *The differential equations* (2.101) *(i.e.* (2.3) *and* (2.22)*) are compatible if and only if the parameters q, p satisfy the following equation*:

$$q_t = \frac{\partial H}{\partial p}, \quad p_t = -\frac{\partial H}{\partial q}, \tag{2.102}$$

where $H = -\operatorname*{Res}\limits_{x=t} b(x)$ is given in (2.20).

This equation is the Hamiltonian form of P_{VI}.

Proof For simplicity, by a gauge transformation $y(x) \rightarrow w(x)y(x)$, $w'(x) = \frac{1}{2}a(x)w(x)$, we consider the case where

$$Ly = \{\partial_x^2 + u(x)\}y = 0, \tag{2.103}$$
$$u(x) = -\frac{1}{4}a(x)^2 - \frac{1}{2}a'(x) + b(x).$$

We will show that

$$q_t = \frac{\partial K}{\partial \mu}, \quad \mu_t = -\frac{\partial K}{\partial q}, \tag{2.104}$$

where

$$\mu := \operatorname*{Res}_{x=q} u(x) = p - \frac{1}{2}\left(\frac{m+n}{q} + \frac{\alpha-1}{q-1} + \frac{\beta-1}{q-t}\right), \tag{2.105}$$

$$K := -\operatorname*{Res}_{x=t} u(x) = H + \frac{\beta-1}{2}\left(\frac{m+n}{t} + \frac{\alpha-1}{t-1} + \frac{1}{t-q}\right). \tag{2.106}$$

Since the transformation $(p, q, H, t) \rightarrow (\mu, q, K, t)$ is symplectic, i.e.

$$dp \wedge dq - dH \wedge dt = d\mu \wedge dq - dK \wedge dt, \tag{2.107}$$

the Eqs. (2.102) and (2.104) are equivalent.

A tedious but straightforward computation shows that for L of (2.103) and B of (2.101) the commutator $[L, B] = LB - BL$ is given by

$$[L, B] = (G'' + 2F')\partial_x + (F'' - 2uG' - u_t - Gu'). \tag{2.108}$$

Thus, the compatibility is given by $F' = -\frac{1}{2}G''$ and

$$u_t + (\frac{1}{2}\partial_x^3 + 2u\partial_x + u')G = 0. \tag{2.109}$$

This gives a system of evolution equations for coefficients of $u(x)$ with respect to the time variable t. To analyze the condition (2.109) further, we put the expansions of $u(x)$, $G(x)$ around $z = x - q = 0$, as

$$u(x) = \sum_{n=-2}^{\infty} u_n z^n, \quad G(x) = \sum_{n=-1}^{\infty} \xi_n z^n. \tag{2.110}$$

Note that $u_{-2} = -\frac{3}{4}$ (due to the exponents at $x = q$ are $\frac{1}{2}, -\frac{3}{2}$), and

$$u_{-1} = \mu, \quad u_0 = -\mu^2, \tag{2.111}$$

due to the non-logarithmic condition (2.19). Hence we have the expansions

$$u_t = -\frac{3}{2}q_t z^{-3} + \mu q_t z^{-2} + \mu_t z^{-1} + \cdots, \tag{2.112}$$

$$(\frac{1}{2}\partial_x^3 + 2u\partial_x + u')G = \frac{1}{2}G''' + \sum_{m=-2}^{\infty}\sum_{n=-1}^{\infty}(m+2n)u_m\xi_n z^{m+n-1}$$

$$= (3 + 4u_{-2})\xi_{-1}z^{-4} + (3u_{-1}\xi_{-1} + 2u_{-2}\xi_0)z^{-3}$$

$$+ (2u_0\xi_{-1} + u_{-1}\xi_0)z^{-2} + (u_1\xi_{-1} - u_{-1}\xi_1 - 2u_{-2}\xi_2)z^{-1} + \cdots.$$

Plugging these into (2.109), we obtain

$$q_t = -2\xi_{-1}\mu + \xi_0 = \operatorname*{Res}_{x=q}\left(\frac{\partial u}{\partial \mu}G\right),$$

$$\mu_t = \frac{3}{2}\xi_2 - \xi_1\mu + \xi_{-1}u_1 = -\operatorname*{Res}_{x=q}\left(\frac{\partial u}{\partial q}G\right). \tag{2.113}$$

In the case of (2.22), i.e. $G(x) = \dfrac{x(x-1)(t-q)}{t(t-1)(x-q)}$, the poles of $\dfrac{\partial u}{\partial \mu}G$ are only at $x = q, t$. Then, by the residue theorem, we have

$$q_t = \operatorname*{Res}_{x=q}\left(\frac{\partial u}{\partial \mu}G\right) = -\operatorname*{Res}_{x=t}\left(\frac{\partial u}{\partial \mu}G\right) = \frac{\partial K}{\partial \mu}, \tag{2.114}$$

$$\mu_t = -\operatorname*{Res}_{x=q}\left(\frac{\partial u}{\partial q}G\right) = \operatorname*{Res}_{x=t}\left(\frac{\partial u}{\partial q}G\right) = -\frac{\partial K}{\partial q}.$$

Thus we obtained (2.104), hence (2.102). Though these equations were derived as necessary conditions for the compatibility, one can show that they are sufficient by looking at the local computation at $x = 0, 1, t, \infty$ also. Namely, if the Eq. (2.104) is satisfied, then we can check that $A := (\text{LHS of (2.109)})$ has no poles on $\mathbb{P}^1$ and vanishes at $x = \infty$, hence $A \equiv 0$. $\qquad\square$

As we have seen above, the linear equations (2.101) are compatible if and only if the P_{VI} Eq. (2.102) is satisfied. In this sense, they are called the *Lax pair* for P_{VI}. Such a formulation of nonlinear equations gives a firm base for the inverse scattering method [1]. The equation $Ly = 0$ in (2.101) is a Fuchsian equation on $\mathbb{P}^1 \setminus \{0, 1, t, \infty\}$, and the equation $By = 0$ describes its deformation. Note that y_t has the same monodromy with y and y_x since the coefficients of B are rational functions of x, hence the deformation is an *isomonodromic deformation*. As a result, the sixth Painlevé equation P_{VI} was derived as the isomonodromic deformation of the equation (2.3). Such an isomonodromic interpretation of the Painlevé equations was first obtained by R. Fuchs [7].

From this point of view, we could foresee the relation between the Padé problem and the Painlevé equation from the beginning, since the monodromy of the functions $P(x)$, $\psi(x)Q(x)$ obviously does not depend on t.

Garnier system

As a generalization of (2.103), consider the Fuchsian differential equation

$$Ly = \{\partial_x^2 + u(x)\}y = 0,$$
$$u(x) = \sum_{i=1}^{n} \left\{ \frac{-\frac{3}{4}}{(x - \lambda_i)^2} + \frac{\mu_i}{x - \lambda_i} \right\} + \sum_{i=1}^{n+3} \left\{ \frac{c_i}{(x - t_i)^2} - \frac{K_i}{x - t_i} \right\}. \tag{2.115}$$

We set the regularity at $x = \infty$ and non-logarithmic conditions at $x = \lambda_i$:

$$u(x) = O(x^{-4}) \quad (x \to \infty),$$
$$u(x) = \frac{-\frac{3}{4}}{(x - \lambda_i)^2} + \frac{\mu_i}{x - \lambda_i} - \mu_i^2 + O((x - \lambda_i)^1) \quad (x \to \lambda_i), \tag{2.116}$$

namely

$$\sum_{i=1}^{n} \lambda_i^k \{\lambda_i \mu_i - \frac{3}{4}(k + 1)\} + \sum_{i=1}^{n+3} t_i^k \{-t_i K_i + c_i(k + 1)\} = 0 \quad (k = 0, \pm 1),$$

$$\mu_j^2 + \sum_{i(\neq j)=1}^{n} \left\{ \frac{-\frac{3}{4}}{(\lambda_j - \lambda_i)^2} + \frac{\mu_i}{\lambda_j - \lambda_i} \right\}$$

$$+ \sum_{i=1}^{n+3} \left\{ \frac{c_i}{(\lambda_j - t_i)^2} - \frac{K_i}{\lambda_j - t_i} \right\} = 0 \quad (j = 1, \ldots, n). \tag{2.117}$$

Solving these relations, one can determine K_i as rational functions of $\{\lambda_i, \mu_i, t_i, c_i\}$.

The Garnier system is defined as the isomonodromic deformation of the equation (2.115). It is given as a rather complicated coupled partial differential equation for $2n$ unknown variables $\{\lambda_i, \mu_i\}$ w.r.t. $(n + 3)$-independent variables $\{t_i\}$, The system is obtained as the compatibility of (2.115) and its deformation equation

$$\{\partial_{t_i} + G_i(x)\partial_x + F_i(x)\}y = 0, \tag{2.118}$$

where

$$F_i(x) = -\frac{1}{2}G_i'(x), \quad G_i(x) = \frac{\Lambda(t_i)}{T'(t_i)} \frac{T(x)}{(x - t_i)\Lambda(x)},$$

$$\Lambda(x) = \prod_{i=1}^{n}(x - \lambda_i), \quad T(x) = \prod_{i=1}^{n+3}(x - t_i). \tag{2.119}$$

Fundamentally, the Garnier system can be described as a multi-time Hamiltonian system (2.59) where the Hamiltonian K_i is the function determined by (2.117).

We do not go into the proof of this here, but we will give a few remarks.

(i) The Hamiltonian equation can be obtained through the same residue argument as the P_{VI} case, where we do not need the explicit expression of K_i.

(ii) Due to the $\mathfrak{sl}_2$ symmetry of $\mathbb{P}^1$, three of the $(n + 3)$ variables $\{t_i\}$ can be fixed to any values (e.g. 0, 1, ∞) and the system has effectively n-independent variables. The standard Garnier system is written in this way (see for example Appendix C of [14]). One recovers the Painlevé VI as the case $n = 1$.

(iii) Though the Hamiltonian K_i is rational function in $\{\lambda_i, \mu_i\}$, it can be transformed into a polynomial Hamiltonian system by certain coordinate transformations [26]. Such a polynomial Hamiltonian system is obtained as the 2×2 case of the matrix Lax form (the Schlesinger system). We will study it in the last chapter (see Sect. 6.2).

Chapter 3
Padé Approximation for
q-Painlevé/Garnier Equations

Abstract In this chapter, we consider the Padé approximation where the function $\psi(x)$ is given by certain infinite products and apply it to the q-difference analog of the Painlevé/Garnier equations. As we will see later, the discrete analogs of the Padé approximation (Padé interpolations) are a more natural setting in which to approach the discrete Painlevé/Garnier equations. Hence the results in this chapter may be considered as a separate case; however, they show that there is a profound interplay between discrete and continuous cases.

3.1 Lax Pair for the q-Garnier Equation

There are various discrete analogs of the Painlevé equations [50] and their geometric classification is given in [51]. The main target of this chapter is the q-P_{VI} *equation* (the q-analog of the Painlevé VI) [19], see (3.29) below. Since the treatment is essentially the same also for its multivariable extensions (*q-Garnier system* [52]), we will start with the general setting corresponding to the q-Garnier system with $2N$ variables. The q-P_{VI} equation corresponds to the $N = 1$ case.

We assume that $q \in \mathbb{C}$, $0 < |q| < 1$, and we use the notation of a q-factorial symbol $(x)_s = (x; q)_s$, namely

$$(x)_\infty = \prod_{i=0}^{\infty} (1 - q^i x), \quad (x)_s = \frac{(x)_\infty}{(xq^s)_\infty}, \tag{3.1}$$

and $(x_1, x_2, \ldots, x_k)_s = (x_1)_s (x_2)_s \ldots (x_k)_s$.

© The Author(s), under exclusive license to Springer Nature Singapore Pte Ltd 2021
H. Nagao and Y. Yamada, *Padé Methods for Painlevé Equations*,
SpringerBriefs in Mathematical Physics,
https://doi.org/10.1007/978-981-16-2998-3_3

For the given function

$$\psi(x) = \prod_{i=1}^{N+1} \frac{(a_i x)_\infty}{(b_i x)_\infty}, \tag{3.2}$$

we consider the Padé approximation

$$\psi(x) = \frac{P(x)}{Q(x)} + O(x^{m+n+1}), \tag{3.3}$$

where $P(x)$, $Q(x)$ are polynomials in x of degree m, n and a_i, b_i are complex parameters.

Using the relation

$$\log(x)_\infty = \sum_{n=0}^\infty \log(1 - xq^n) = -\sum_{n=0}^\infty \sum_{k=1}^\infty \frac{(xq^n)^k}{k} = -\sum_{k=1}^\infty \frac{x^k}{k(1-q^k)}, \tag{3.4}$$

the function $\psi(x)$ can be written as

$$\psi(x) = \exp\left(\sum_{k=1}^\infty \sum_{s=1}^{N+1} \frac{b_s^k - a_s^k}{k(1-q^k)} x^k \right). \tag{3.5}$$

q-binomial theorem

The following is well known and fundamental in q-analysis.
Lemma 3.1 We have

$$\frac{(ax)_\infty}{(x)_\infty} = \sum_{n=0}^\infty \frac{(a)_n}{(q)_n} x^n. \tag{3.6}$$

Proof Due to the relation $(qz)_\infty = \frac{1}{(1-z)}(z)_\infty$, the LHS $y(x) = \frac{(ax)_\infty}{(x)_\infty}$ satisfies the q-difference equation

$$(1 - ax)y(qx) = (1 - x)y(x). \tag{3.7}$$

Solve this as a power series in x, with the initial condition $y(0) = 1$. Then we obtain the RHS. □
The formula (3.6) is called the q-binomial theorem, because we have

$$\frac{(q^\alpha x)_\infty}{(x)_\infty} \to (1-x)^{-\alpha} \quad (q \to 1). \tag{3.8}$$

Putting $a_i = b_i q^{-\alpha_i}$ and taking a limit $q \to 1$, we have $\psi(x) = \prod_{i=1}^{N+1}(1 - b_i x)^{\alpha_i}$, which has the form $\psi(x)$ (2.54) for the differential Garnier systems. So one can expect the Padé problem for $\psi(x)$ (3.2) to be related to a certain q-analog of the Garnier system.

Let $P(x)$ and $Q(x)$ be the polynomials of degree m and n determined by the Padé condition (1.10) normalized as $P(0) = Q(0) = 1$. For the parameter shift, we take $T = T_{a_1} T_{b_1}$ such that

$$T(f) = f\Big|_{a_1 \to q a_1, b_1 \to q b_1}. \tag{3.9}$$

We also use the following *up/down shift notation*:

$$\overline{f} = T(f), \quad \underline{f} = T^{-1}(f). \tag{3.10}$$

Let us consider the equations for $y(x)$ satisfied by $y(x) = P(x)$ and $y(x) = \psi(x)Q(x)$. The main equation is the three-term relation L_1 between $y(qx)$, $y(x)$, $y(\frac{x}{q})$; however, it is a little difficult to determine the explicit form. So we will begin with the contiguity type relations, from which the L_1 equation will be considered in the next section.

Theorem 3.1 *The linear difference relation L_2 between $y(x)$, $y(qz)$, $\overline{y}(x)$, and the relation L_3 between $y(x)$, $\overline{y}(x)$, $\overline{y}(x/q)$ satisfied by the functions $y(x) = P(x)$ and $y(x) = \psi(x)Q(x)$ are given as follows:*[1]

$$L_2 \ : (g_0)_1 F(x)\overline{y}(x) - A_1(x)y(qx) + (b_1 x)_1 G(x) y(x) = 0, \tag{3.11}$$

$$L_3 \ : (\frac{g_0}{c})_1 \overline{F}(\frac{x}{q})y(x) + \frac{1}{c}(a_1 x)_1 G(\frac{x}{q})\overline{y}(x) - B_1(\frac{x}{q})\overline{y}(\frac{x}{q}) = 0, \tag{3.12}$$

where $c = q^{m+n+1}$. $A(x)$, $B(x)$, $A_i(x)$, $B_i(x)$, $F(x)$, $G(x)$ are the following polynomials in x:

$$A(x) = \prod_{i=1}^{N+1}(a_i x)_1, \quad B(x) = \prod_{i=1}^{N+1}(b_i x)_1, \quad F(x) = 1 + \sum_{i=1}^{N} f_i x^i, \tag{3.13}$$

$$A_i(x) = \frac{A(x)}{(a_i x)_1}, \quad B_i(x) = \frac{B(x)}{(b_i x)_1}, \quad G(x) = \sum_{i=0}^{N-1} g_i x^i, \tag{3.14}$$

and $f_1, \ldots, f_N, g_0, \ldots, g_{N-1}$ are some constants which play the role of dependent variables of the q-Garnier system.

Proof The derivation of the equations (3.11), (3.12) is similar to the previous examples and we will show the outline of the computation. In this proof we use another *up/down-shift notation* such as

---

[1] We use the notation $(z)_1 = (1 - z)$. Though this is a non-standard use of the q-factorial symbol (3.1), it is sometimes useful to shorten the expressions.

$$f^x = f\Big|_{x \to qx}, \quad f_x = f\Big|_{x \to x/q}. \tag{3.15}$$

By the definition, L_2 is given by

$$L_2 : \begin{vmatrix} y & y^x & \overline{y} \\ \mathbf{u} & \mathbf{u}^x & \overline{\mathbf{u}} \end{vmatrix} = D_1 \overline{y} - D_2 y^x + D_3 y = 0, \tag{3.16}$$

where D_1, D_2, D_3 are *Casorati determinants*:

$$D_1 = |\mathbf{u}, \mathbf{u}^x|, \quad D_2 = |\mathbf{u}, \overline{\mathbf{u}}|, \quad D_3 = |\mathbf{u}^x, \overline{\mathbf{u}}|, \quad \mathbf{u} = \begin{bmatrix} P \\ \psi Q \end{bmatrix}. \tag{3.17}$$

Noting that $\frac{\psi^x}{\psi} = \frac{B}{A}$ we have

$$\begin{aligned}
D_1 &= \begin{vmatrix} P & P^x \\ \psi Q & \psi^x Q^x \end{vmatrix} = \psi \begin{vmatrix} P & P^x \\ Q & \frac{B}{A} Q^x \end{vmatrix} \\
&= \frac{\psi}{A} \{ B P Q^x - A P^x Q \} \\
&= \frac{\psi x^{m+n+1}}{A} x^{[N]} = \frac{\psi x^{m+n+1}}{A} w_0 F(x),
\end{aligned} \tag{3.18}$$

where $x^{[N]}$ represents a polynomial in x of degree N, and w_0 is a constant. Similarly from $\frac{\overline{\psi}}{\psi} = \frac{(b_1 x)_1}{(a_1 x)_1}$, we have

$$\begin{aligned}
D_2 &= \begin{vmatrix} P & \overline{P} \\ \psi Q & \overline{\psi Q} \end{vmatrix} = \psi \begin{vmatrix} P & \overline{P} \\ Q & \frac{(b_1 x)_1}{(a_1 x)_1} \overline{Q} \end{vmatrix} \\
&= \frac{\psi}{(a_1 x)_1} \{ (b_1 x)_1 P \overline{Q} - (a_1 x)_1 \overline{P} Q \} \\
&= \frac{\psi x^{m+n+1}}{(a_1 x)_1} x^{[0]} = \frac{\psi A_1 x^{m+n+1}}{A} w_1.
\end{aligned} \tag{3.19}$$

We also have

$$\begin{aligned}
D_3 &= \begin{vmatrix} P^x & \overline{P} \\ (\psi Q)^x & \overline{\psi Q} \end{vmatrix} = \psi \begin{vmatrix} P^x & \overline{P} \\ \frac{B}{A} Q^x & \frac{(b_1 x)_1}{(a_1 x)_1} \overline{Q} \end{vmatrix} \\
&= \frac{\psi (b_1 x)_1}{A} \{ A_1 P^x \overline{Q} - B_1 \overline{P} Q^x \} \\
&= \frac{\psi (b_1 x)_1 x^{m+n+1}}{A} x^{[N-1]} = \frac{\psi (b_1 x)_1 x^{m+n+1}}{A} w_1 G(x),
\end{aligned} \tag{3.20}$$

with some constant w_1. Then the L_2 equation has the form

$$\frac{w_0}{w_1} F(x)\overline{y}(x) - A_1(x)y(qx) + (b_1 x)_1 G(x)y(x) = 0. \qquad (3.21)$$

We see that $\frac{w_0}{w_1} = (g_0)_1$ from the solution $y(x) = P(x)$ with $P(0) = 1$, hence we obtain (3.11).

For L_3 equation, its x-up shift L_3^x is given by

$$L_3^x : \begin{vmatrix} y^x & \overline{y}^x & \overline{y} \\ \mathbf{u}^x & \overline{\mathbf{u}}^x & \overline{\mathbf{u}} \end{vmatrix} = \overline{D_1}y^x + D_3\overline{y}^x - D_2^x\overline{y} = 0, \qquad (3.22)$$

where the coefficients are

$$\overline{D_1} = \overline{\left\{ w_0 \frac{\psi x^{m+n+1}}{A} F \right\}} = \frac{(b_1 x)_1}{(a_1 x)_1} \overline{w_0} \frac{\psi x^{m+n+1}}{(a_1 q x)_1 A_1} \overline{F}, \qquad (3.23)$$

$$D_3 = w_1 \frac{\psi x^{m+n+1}}{A} (b_1 x)_1 G,$$

$$D_2^x = \left\{ w_1 \frac{\psi x^{m+n+1}}{(a_1 x)_1} \right\}^x = w_1 \frac{B}{A} \frac{\psi (qx)^{m+n+1}}{(a_1 q x)_1}.$$

Hence L_3^x takes the form

$$\frac{\overline{w_0}}{w_1} \overline{F} y(qx) - (a_1 q x)_1 G(x)\overline{y}(qx) + B_1(x)q^{m+n+1}\overline{y}(x) = 0, \qquad (3.24)$$

and, again from the solution $y(x) = P(x) = 1 + O(x)$, we have $\frac{\overline{w_0}}{w_1} = g_0 - q^{m+n+1}$ and the expression (3.12) for L_3. □

In the discrete case also, we will use the on/off-Padé situations (analogous to continuous case, §2.6) according to our purpose. And now, we switch to the off-Padé situation. Then we have the following:

Proposition 3.1 *In the off-Padé situation, the compatibility of the equations L_2, L_3 in (3.11), (3.12) gives the following relations ($c = q^{m+n+1}$) :*

$$A_1(x)B_1(x) - \frac{1}{c}(a_1 x, b_1 x)_1 G(x)\underline{G}(x) = 0 \quad \text{for} \quad F(x) = 0, \qquad (3.25)$$

$$A_1(x)B_1(x) - (g_0, \frac{g_0}{c})_1 F(x)\overline{F}(x) = 0 \quad \text{for} \quad G(x) = 0, \qquad (3.26)$$

$$(g_0, \frac{g_0}{c})_1 f_N \overline{f_N} = \frac{q a_1 b_1}{c} \left(g_{N-1} + \frac{q^n}{a_1} \prod_{i=2}^{N+1}(-b_i) \right)\left(g_{N-1} + \frac{q^m}{b_1} \prod_{i=2}^{N+1}(-a_i) \right). \qquad (3.27)$$

Proof Equation (3.25) is obtained from $L_2(x)$ and $\underline{L_3}(x)$, and (3.26) is from $L_2(x)$ and $L_3(qx)$. Considering the solution $y(x) = P(x) = Cx^m + \text{lower}$, the highest degree terms of $L_2(x)$, $L_3(qx)$ give

$$(g_0)_1 f_N \overline{C} = \left(b_1 g_{N-1} + q^m \prod_{i=2}^{N+1}(-a_i)\right)C,$$

$$(\frac{g_0}{c})_1 \overline{f_N} C = \frac{q}{c}\left(a_1 g_{N-1} + q^n \prod_{i=2}^{N+1}(-b_i)\right)\overline{C}, \qquad (3.28)$$

hence we obtain (3.27).   □

One can prove [41] that the relations (3.25)–(3.27) are sufficient for the compatibility of L_2, L_3 and these relations can be regarded as evolution equations of the q-Garnier system [52] along the direction $T = T_{a_1} T_{b_1}$. The equations are birational evolution equations for $2N$ variables (the coefficients of $F(x) = 1 + f_1 x + \cdots + f_N x^N$ and $G(x) = g_0 + \cdots + g_{N-1} x^{N-1}$). Since their explicit forms are rather complicated in general, we will show simpler cases in the next example.

Example

For $N = 1$, in terms of the variables (f, g) such that $F(x) = 1 - fx$ and $G(x) = g$, we have the equations

$$f\overline{f} = \frac{(g - q^m \frac{a_2}{b_1})(g - q^n \frac{b_2}{a_1})}{(g - 1)(g - q^{m+n+1})} q a_1 b_1, \qquad (3.29)$$

$$g\underline{g} = \frac{(f - a_2)(f - b_2)}{(f - a_1)(f - b_1)} q^{m+n+1},$$

together with $\overline{a_1} = q a_1, \overline{b_1} = q b_1$. This system is known as the Jimbo–Sakai q-P_{VI} [19] (the q-analog of P_{VI}), where m, n are not restricted to integers.

For $N = 2$, the equation is a system with four variables in general. Under a suitable specialization, one can reduce it to a system of two variables [53] (see also [41][2]). To do this, we put a constraint $q^m a_1 a_2 a_3 = q^n b_1 b_2 b_3$ on the parameters, then from (3.18) the degree of the polynomial $F(x)$ is reduced from 2 to 1 (hence $f_2 = 0$). We can also put $g_1 = -\frac{q^n b_2 b_3}{a_1} = -\frac{q^m a_2 a_3}{b_1}$ consistently with (3.27). Then, in terms of variables $f = -f_1$ and $g = -\frac{g_0}{g_1}$, the system can be written as

$$(fg - 1)(f\underline{g} - 1) = \frac{(1 - \frac{f}{a_2})(1 - \frac{f}{a_3})(1 - \frac{f}{b_2})(1 - \frac{f}{b_3})}{(1 - \frac{f}{a_1})(1 - \frac{f}{b_1})}, \qquad (3.30)$$

$$(fg - 1)(\overline{f}g - 1) = \frac{(1 - a_2 g)(1 - a_3 g)(1 - b_2 g)(1 - b_3 g)}{(1 - a_4 g)(1 - b_4 g)},$$

---

[2] In the second of equations (2.23) in [41], $(f\overline{g} - 1)$ should read $(\overline{f}g - 1)$.

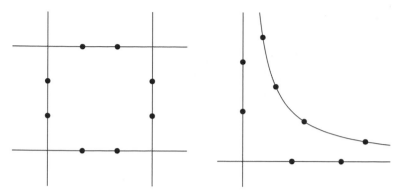

Fig. 3.1 The singular point configuration for q-P_{VI} (left) and q-$P(E_6^{(1)})$ (right)

where $a_4 = \frac{q^n b_2 b_3}{a_1}$, $b_4 = \frac{q^{-n-1} a_2 a_3}{b_1}$ $(\overline{a_1} = q a_1, \overline{b_1} = q b_1)$. This system is known as the q-*Painlevé equation of type* $E_6^{(1)}$. The Padé approximation problem related to this equation was first studied in [16] (see also [37]).

A convenient way to specify the (discrete) Painlevé equations is to consider the configuration of the singular points (i.e. the points where the birational mapping has indeterminacy). The configuration for (3.29) is on the four lines $f = 0$, $f = \infty$, $g = 0$, $g = \infty$, and that for (3.30) is on the two lines $f = 0$, $g = 0$ and a curve $fg = 1$ (see Fig. 3.1).

3.2 The L_1 Equation

From (3.18) and its shift $x \to x/q$, we see

$$|\mathbf{u}, \mathbf{u}^x| = \frac{\psi(x)}{A(x)} w_0 x^{m+n+1} F(x), \quad |\mathbf{u}_x, \mathbf{u}| = \frac{\psi(x)}{B(\frac{x}{q})} w_0 (\frac{x}{q})^{m+n+1} F(\frac{x}{q}). \quad (3.31)$$

Also it is not difficult to see

$$|\mathbf{u}_x, \mathbf{u}^x| = \frac{\psi(x)}{A(x)B(\frac{x}{q})} x^{m+n+1} x^{[2N+1]}. \quad (3.32)$$

Thus the L_1 equation takes the form

$$L_1 : A(x)F(\frac{x}{q})y(qx) - M(x)y(x) + q^{m+n+1} B(\frac{x}{q})F(x)y(\frac{x}{q}) = 0, \quad (3.33)$$

where $M(x)$ is a polynomial of degree $2N + 1$. A characterization of $M(x)$ is as follows: (i) Since the multiplicative exponents of the equation L_1 are $1, q^{m+n+1}$ (at $x = 0$) and $q^m, q^n \prod_{i=1}^{N+1} \frac{b_i}{a_i}$ (at $x = \infty$),[3] the coefficients of the lowest and the highest terms are $C_0 = 1 + q^{m+n+1}$ and $C_{2N+1} = q^{-N} f_N \Big\{ q^m \prod_{i=1}^{N+1}(-a_i) + q^n \prod_{i=1}^{N+1}(-b_i) \Big\}$. (ii) For the roots λ_i of the polynomial $F(x)$ and variables μ_i defined by

$$\mu_i = \frac{y(q\lambda_i)}{y(\lambda_i)} \quad (i = 1, \ldots, N), \tag{3.34}$$

we have the relations

$$\mu_i = \frac{M(\lambda_i)}{A(\lambda_i)F(\frac{\lambda_i}{q})}, \quad \frac{1}{\mu_i} = q^{-m-n-1} \frac{M(q\lambda_i)}{B(\lambda_i)F(q\lambda_i)}. \tag{3.35}$$

The relations ensure the consistency of the L_1 equation at $x = \lambda_i$ and $x = q\lambda_i$ (a certain analog of the non-logarithmic condition; see the explanation after Remark 3.1).

An explicit expression for the coefficient $M(x)$ is as follows. From L_2 (3.11) we have

$$\overline{y}(x) = \frac{1}{(g_0)_1 F(x)} \{A_1(x)y(qx) - (b_1 x)_1 G(x)y(x)\}. \tag{3.36}$$

Using this, eliminate $\overline{y}(x)$ and $\overline{y}(\frac{x}{q})$ from L_3 (3.12); we arrive at the L_1 equation with

$$M(x) = c \frac{(g_0, \frac{g_0}{c})_1 F(x)F(\frac{x}{q})\overline{F}(x) - A_1(\frac{x}{q})B_1(\frac{x}{q})F(x)}{G(\frac{x}{q})} \\ - (a_1 x, b_1 x)_1 F(\frac{x}{q})G(\frac{x}{q}). \tag{3.37}$$

Note that this $M(x)$ is a polynomial in x due to the relation (3.26).

Remark 3.1 The variables $\{f_i, g_i\}$ and $\{\lambda_i, \mu_i\}$ are related by $F(\lambda_i) = 0$ and $A_1(\lambda_i)\mu_i = (b_1\lambda_i)_1 G(\lambda_i)$ $(i = 1, \ldots, N)$. The variables $\{\lambda_i, \mu_i\}$ have simple interpretation as the canonical coordinates defined through the Sklyanin's *magic recipe* (see [54, 55] for example). Contrary to the variables $\{f_i, g_i\}$, the evolution equation in terms of $\{\lambda_i, \mu_i\}$ are not birational since we need roots λ_i of the algebraic equation $F(x) = 0$.

q-analog of the non-logarithmic condition

---

[3] The constraint $q^m a_1 a_2 a_3 = q^n b_1 b_2 b_3$ for the reduction in the previous example can be interpreted as a degeneration condition of the exponents at $x = \infty$.

Consider a q-difference equation of order n for unknown function $y(x)$

$$a_0(x)y(x) + a_1(x)y(qx) + \cdots + a_n(x)y(q^n x) = 0, \qquad (3.38)$$

where the coefficients $a_i(x)$ are assumed to be polynomials for simplicity.

Since $x = 0, \infty$ are fixed points of the q-shift operation $x \mapsto qx$, the study of the local solutions of (3.38) has a different nature for $x = 0, \infty$ and other cases.

Solutions around $x = 0, \infty$ We will concentrate on the case $x = 0$, since the case $x = \infty$ can be treated similarly using the local coordinate x^{-1}. To find the power series solutions around $x = 0$, we put

$$y_i(x) = x^{\rho_i} \sum_{k=0}^{\infty} c_{i,k} x^k \quad (c_{i,0} = 1). \qquad (3.39)$$

Then the exponents ρ_i are determined by the equation

$$a_0(0) + a_1(0)t + \cdots + a_n(0)t^n = 0, \quad t = q^{\rho_i}. \qquad (3.40)$$

If the exponents ρ_i are generic, the series solutions y_i are unique and form a fundamental solution. Similar to the differential case, when some of the exponents have integer differences, there are two cases: (i) a special case where we still have power series solutions, (ii) a generic case where one should add logarithmic terms to some solutions.

For example, consider the following equation with exponents $\rho_i = 0, 2$:

$$(q^2 + a_1 x + a_2 x^2 + \cdots)y(x) + \left\{ -(1 + q^2) + b_1 x + b_2 x^2 + \cdots \right\} y(qx)$$
$$\qquad (3.41)$$
$$+ (1 + c_1 x + c_2 x^2 + \cdots)y(q^2 x) = 0.$$

We look for a solution corresponding to $\rho_1 = 0$ of the form $y_1 = 1 + u_1 x + u_2 x^2 + \cdots$. Substituting this into (3.41), we have the equations for the coefficients $u_1, u_2, \ldots$ as

$$q(1 - q)^2 u_1 = a_1 + b_1 + c_1,$$
$$0 \cdot u_2 = a_2 + b_2 + c_2 + (a_1 + q b_1 + q^2 c_1)u_1, \quad \ldots . \qquad (3.42)$$

If the RHS of the 2nd equation is zero (case (i)), the coefficient u_2 is free and we have two series solutions corresponding to $\rho_i = 0, 2$, otherwise (case (ii)) the solution has a logarithmic term.

Solutions around $x = c(\neq 0, \infty)$ We rewrite (3.38) as

$$y(q^n x) = -\frac{1}{a_n(x)} \{a_0(x)y(x) + \cdots + a_{n-1}(x)y_{n-1}(q^{n-1}x)\}. \qquad (3.43)$$

Using this recursively, $y(q^k x)$ ($k \geq n$) can be expressed in terms of $y(x)$, $y(qx)$, $\cdots$, $y(q^{n-1}x)$. Since the denominator has factors $a_n(x)a_n(qx)\cdots a_n(q^{k-n}x)$, the solution will have poles $c, qc, q^2c, \cdots$ for the zeros c of the coefficient $a_n(x)$. Such an infinite sequence of poles can be considered as a singularity (discrete analog of the branch cut). Similarly, the zeros of $a_0(x)$ also could be a source of singularities (for $y(q^k x)$ ($k << 0$)). However, in a certain fortunate situation, the coefficients $a_0(x)$ and/or $a_n(x)$ can have zeros which do not cause the infinite sequences of poles for solutions. For instance, in the 2nd-order case, we have

$$y(q^2 x) = -\frac{1}{a_2}(a_0 y + a_1 \bar{y}), \quad y(q^3 x) = -\frac{1}{\bar{a}_2}\{\overline{a_0 y} - \frac{\overline{a_1}}{a_2}(a_0 y + a_1 \bar{y})\}, \quad (3.44)$$

where $\bar{y} = y(qx)$ etc. Then a possible zero of $\bar{a}_2 = a_2(qx)$ at $x = c/q$ in the denominator of $y(q^3 x)$ can be cancelled if both a_0 and $a_1 \overline{a_1} - \overline{a_0} a_2$ also vanish at $x = c/q$. We will call such a case an "apparent singularity" (with non-logarithmic property) by abuse of terminology. For example, the point $x = c$ is an apparent singularity of

$$\tilde{a}_0(x)(qx - c)y(x) + a_1(x)y(qx) + \tilde{a}_2(x)(x - c)y(q^2 x) = 0, \quad (3.45)$$

if the coefficients satisfy the relation $\tilde{a}_0(c)\tilde{a}_2(c/q)(1 - q)^2 c^2 = q a_1(c)a_1(c/q)$. Note that one can derive this relation easily as the consistency of the two equations for $y(qc)$ and $y(q^2 c)$ obtained from (3.45) by setting $x = c$ and $x = qc$. The apparent singularities in difference equation play an important role also in [58] (see also [43]).

3.3 Special Solutions

We derive the explicit forms of the polynomials $F(x)$, $G(x)$ appearing in the Casorati determinants D_1 and D_3 (3.23). They are interpreted as the special solutions for the q-Garnier system (3.25)–(3.27). To do this, we use the formulas of $P(x)$ and $Q(x)$ in (2.33) and (2.69).

Lemma 3.2 *We have the following expression for special values of the polynomials $P(x)$ and $Q(x)$:*

$$P\left(\frac{1}{a_s}\right) = \left(\frac{1}{a_s}\right)^m T_{a_s}(\tau_{m,n+1}), \quad Q\left(\frac{q}{a_s}\right) = \left(-\frac{q}{a_s}\right)^n T_{a_s}^{-1}(\tau_{m+1,n}),$$

$$P\left(\frac{q}{b_s}\right) = \left(\frac{q}{b_s}\right)^m T_{b_s}^{-1}(\tau_{m,n+1}), \quad Q\left(\frac{1}{b_s}\right) = \left(-\frac{1}{b_s}\right)^n T_{b_s}(\tau_{m+1,n}),$$

$$(3.46)$$

for $s = 1, \ldots, N + 1$, where $\tau_{m,n} = s_{(m^n)}$.

Proof We apply the general results in previous sections to $\psi(x) = \sum_{k=0}^{\infty} p_k x^k$ in (3.2), By noting that the shifts $p_i \to \sum_{j=0}^{i} a^j p_{i-j}$ and $p_i \to p_i - a p_{i-1}$ correspond to $\psi(x) \to (1 - ax)^{-1} \psi(x)$ and $\psi(x) \to (1 - ax) \psi(x)$ respectively, we have

$$T_{a_s}^{-1}(p_i) = V(\frac{q}{a_s})p_i = p_i - \frac{1}{q}a_s p_{i-1}, \quad T_{b_s}(p_i) = V(\frac{1}{b_s})p_i = p_i - b_s p_{i-1},$$

$$T_{a_s}(p_i) = V^*(\frac{q}{a_s})p_i = \sum_{j=0}^{i} a_s^j p_{i-j}, \quad T_{b_s}^{-1}(p_i) = V^*(\frac{q}{b_s})p_i = \sum_{j=0}^{i} (\frac{b_s}{q})^j p_{i-j},$$

(3.47)

for $s = 1, \ldots, N + 1$. Thanks to these formulas and the formula (2.69), we obtain the expressions (3.46). $\qquad \square$

Proposition 3.2 *For the polynomial $F(x)$, we have*

$$C_0 F(\frac{1}{a_i}) = q^n a_i B(\frac{1}{a_i}) T_{a_i}(\tau_{m,n+1}) T_{a_i}^{-1}(\tau_{m+1,n}),$$

(3.48)

$$C_0 F(\frac{1}{b_i}) = -q^m b_i A(\frac{1}{b_i}) T_{b_i}^{-1}(\tau_{m,n+1}) T_{b_i}(\tau_{m+1,n}),$$

(3.49)

$$C_0 f_N = (q^n \prod_{i=1}^{N+1}(-b_i) - q^m \prod_{i=1}^{N+1}(-a_i)) \tau_{m,n+1} \tau_{m+1,n},$$

(3.50)

$$C_0 = (q^{m+n+1} - 1) \tau_{m,n} \tau_{m+1,n+1},$$

(3.51)

where $i = 1, \ldots, N + 1$ and $C_0 = (-1)^n w_0$. Also, for $G(x)$, we have

$$C_1 G(\frac{1}{a_i}) = -q^n a_i B_1(\frac{1}{a_i}) T_{a_i}(\bar{\tau}_{m,n+1}) T_{a_i}^{-1}(\tau_{m+1,n}),$$

(3.52)

$$C_1 G(\frac{1}{b_i}) = q^m b_i A_1(\frac{1}{b_i}) T_{b_i}^{-1}(\tau_{m,n+1}) T_{b_i}(\bar{\tau}_{m+1,n}),$$

(3.53)

$$C_1 g_{N-1} = q^m \prod_{i=2}^{N+1}(-a_i) \bar{\tau}_{m,n+1} \tau_{m+1,n} - q^n \prod_{i=2}^{N+1}(-b_i) \tau_{m,n+1} \bar{\tau}_{m+1,n},$$

(3.54)

$$C_1 g_0 = q^{m+n+1} \tau_{m+1,n+1} \bar{\tau}_{m,n} - \bar{\tau}_{m+1,n+1} \tau_{m,n},$$

(3.55)

where $i = 2, \ldots, N + 1$ and $C_1 = (-1)^n w_1$. We also have

$$C_1 = (a_1 - b_1) T_{a_1}(\tau_{m+1,n}) T_{b_1}(\tau_{m,n+1})$$

(3.56)

$$= -b_1 \tau_{m,n+1} \bar{\tau}_{m+1,n} + a_1 \bar{\tau}_{m,n+1} \tau_{m+1,n} = \tau_{m+1,n+1} \bar{\tau}_{m,n} - \bar{\tau}_{m+1,n+1} \tau_{m,n}.$$

Proof From (3.18), we have

$$B(x)P(x)Q(qx) - A(x)P(qx)Q(x) = x^{m+n+1} w_0 F(x).$$

(3.57)

Putting $x = \frac{1}{a_i}, \frac{1}{b_j}$, we get

$$w_0 F(\frac{1}{a_i}) = a_i^{m+n+1} B(\frac{1}{a_i}) P(\frac{1}{a_i}) Q(\frac{q}{a_i}),　　　　　　(3.58)$$

$$w_0 F(\frac{1}{b_i}) = b_i^{m+n+1} A(\frac{1}{b_i}) P(\frac{q}{b_i}) Q(\frac{1}{b_i}).　　　　　　(3.59)$$

Then substituting (3.46) we obtain (3.48), (3.49). Similarly, considering the limits $x \to \infty$ and $x \to 0$ in (3.57), we have (3.50), (3.51). In the case of $x \to 0$, we used the relation

$$P(x) - \psi(x) Q(x) = (-1)^n \tau_{m+1,n+1} x^{m+n+1} + \text{higher order},　　　(3.60)$$

which follows from (2.39). In a similar way, from (3.20) i.e.

$$A_1(x) P(qx) \overline{Q}(x) - B_1(x) \overline{P}(x) Q(qx) = x^{m+n+1} w_1 G(x),　　　(3.61)$$

we obtain (3.52)–(3.55). Finally, from (3.19) i.e.

$$(b_1 x)_1 P(x) \overline{Q}(x) - (a_1 x)_1 \overline{P}(x) Q(x) = x^{m+n+1} w_1,　　　　(3.62)$$

we obtain (3.56).　　　　　　　　　　　　　　　　　　　　　　　　　　　□

Lemma 3.3 *We have*

$$p_k = \frac{b_{N+1}^k \left(\frac{a_{N+1}}{b_{N+1}} \right)_k}{(q)_k} \varphi_D \left(q^{-k}, \frac{a_1}{b_1}, \ldots, \frac{a_N}{b_N}, q^{-k+1} \frac{b_{N+1}}{a_{N+1}}; \frac{qb_1}{a_{N+1}}, \ldots, \frac{qb_N}{a_{N+1}} \right),$$
$$(3.63)$$

where φ_D is the q-Appell–Lauricella function [10]

$$\varphi_D(\alpha, \beta_1, \ldots, \beta_N, \gamma; z_1, \ldots, z_N) = \sum_{m_i \geq 0} \frac{(\alpha)_{|m|} (\beta_1)_{m_1} \cdots (\beta_N)_{m_N}}{(\gamma)_{|m|} (q)_{m_1} \cdots (q)_{m_N}} z_1^{m_1} \cdots z_N^{m_N},$$
$$(3.64)$$

and $|m| = m_1 + \ldots + m_N$.

Proof By the definition of $\psi(x)$ (3.2) and the q-binomial theorem (3.6), we have

$$\psi(x) = \sum_{m_i \geq 0} \prod_{i=1}^{N+1} \left\{ \frac{(\frac{a_i}{b_i})_{m_i}}{(q)_{m_i}} b_i^{m_i} x^{m_i} \right\},$$
$$p_k = \sum_{\sum_{i=1}^{N+1} m_i = k} \prod_{i=1}^{N+1} \left\{ \frac{(\frac{a_i}{b_i})_{m_i}}{(q)_{m_i}} b_i^{m_i} \right\}.　　　　　(3.65)$$

Putting $m_{N+1} = k - |m|$ and using a relation

$$\frac{(a)_{k-l}}{(b)_{k-l}} = \frac{(q^{-k+1}/b)_l}{(q^{-k+1}/a)_l} \frac{b^l}{a^l} \frac{(a)_k}{(b)_k},　　　　　　(3.66)$$

we get (3.63). □

Example

For the $N = 1$ case, the following expressions

$$f = -\frac{a_1 a_2 q^m - b_1 b_2 q^n}{1 - q^{m+n+1}} \frac{\tau_{m+1,n} \tau_{m,n+1}}{\tau_{m,n} \tau_{m+1,n+1}}, \tag{3.67}$$

$$g = -q^{mn+m+n} \frac{a_2 - b_2}{a_1 - b_1} \frac{T_{a_2}^{-1}(\tau_{m+1,n}) T_{b_2}^{-1}(\tau_{m,n+1})}{T_{a_1}(\tau_{m,n+1}) T_{b_1}(\tau_{m+1,n})}$$

give a special solution of the q-P_{VI} equation (3.29). In this case, p_k in (3.63), (3.64) is given by Heine's q-hypergeometric function $_2\varphi_1$. Similarly, we have special solutions for (3.30) in terms of the φ_D (3.64) in two variables (or equivalently generalized q-hyper geometric function $_3\varphi_2$) [16, 37].

More bilinear relations

Besides the bilinear relations for the general Schur functions, we can derive another kind of bilinear relations satisfied by the τ-functions specialized as (3.5). Note that for a general degree n polynomial $f(x)$, its $n + 1$ values $f(x_i)$ are related by

$$\sum_{i=0}^{n} (-1)^i \Delta(x_0, x_1, \ldots, \widehat{x_i}, \ldots, x_n) f(x_i) = 0, \tag{3.68}$$

where Δ is the difference product $\prod_{i<j}(x_i - x_j)$ and the symbol $\widehat{}$ represents the deletion of the variable. We apply this to the special values (3.48), (3.49), for the polynomial $F(x)$ of degree N. For the index sets $I, J \subset U = \{1, 2, \ldots, N + 1\}$, such that $|I| + |J| = N + 2$, then from the relation between $F(\frac{1}{a_i})$ ($i \in I$) and $F(\frac{1}{b_j})$ ($j \in J$), we have

$$q^m \sum_{i \in I} \frac{\prod_{j \in J^c}(a_i - b_j)}{\prod_{j \in I \setminus \{i\}}(a_i - a_j)} T_{a_i}(\tau_{m,n+1}) T_{a_i}^{-1}(\tau_{m+1,n}) \tag{3.69}$$

$$= q^n \sum_{j \in J} \frac{\prod_{i \in I^c}(b_j - a_i)}{\prod_{i \in J \setminus \{j\}}(b_j - b_i)} T_{b_j}^{-1}(\tau_{m,n+1}) T_{b_j}(\tau_{m+1,n}),$$

where $I^c = U \setminus I, J^c = U \setminus J$. Similarly, for the index sets $I, J \subset U' = \{2, \ldots, N + 1\}$ such that $|I| + |J| = N + 1$, from the relation between $G(\frac{1}{a_i})$ ($i \in I$) and $G(\frac{1}{b_j})$ ($j \in J$) given in (3.52), (3.53), we have

$$q^m \sum_{i \in I} \frac{\prod_{j \in J^c}(a_i - b_j)}{\prod_{j \in I \setminus \{i\}}(a_i - a_j)} T_{a_i}(\overline{\tau_{m,n+1}}) T_{a_i}^{-1}(\tau_{m+1,n}) \tag{3.70}$$

$$= q^n \sum_{j \in J} \frac{\prod_{i \in I^c}(b_j - a_i)}{\prod_{i \in J \setminus \{j\}}(b_j - b_i)} T_{b_j}^{-1}(\tau_{m,n+1}) T_{b_j}(\overline{\tau_{m+1,n}}),$$

where $I^c = U' \setminus I$, $J^c = U' \setminus J$ and $\bar{x} = T_{a_1} T_{b_1}(x)$.

Limit from q-P_{VI} to P_{VI}

The functions $\psi(x)$ in (3.2) (with $N = 1$) and in (2.2) are related as

$$\prod_{i=1}^{2} \frac{(a_i x)_\infty}{(b_i x)_\infty} \to (1 - x)^\alpha (1 - \frac{x}{t})^\beta \quad (\epsilon \to 0), \tag{3.71}$$

by the parametrization

$$a_1 = t^{-1} e^{\epsilon\beta}, \quad b_1 = t^{-1}, \quad a_2 = e^{\epsilon\alpha}, \quad b_2 = 1, \quad q = e^{-\epsilon}, \tag{3.72}$$

where the shift direction is $T(t) = e^\epsilon t$. Hence, it is natural to expect that the resulting Painlevé equations q-P_{VI} and P_{VI}, together with their Lax pairs and special solutions, can be related in a suitable limit. In order to complete this limiting procedure, we should know the relation between the variables (f, g) for q-P_{VI} and (q, p) for P_{VI}. Here we use $q = e^{-\epsilon}$ for the q-shift parameter in order to avoid the confusion with variable q. This can be easily found by using the results from Padé as follows. First, we note that the interpolants $P(x)$, $Q(x)$ in both Padé problems are directly related by the limit through (3.72). This means that the resulting Lax pairs are also related by this limit without any change on x and $y(x)$. In terms of $y(x)$, both the variables f and q are defined as the position of the apparent singularity of the Lax equations. Thus we find

$$f^{-1} = q + O(\epsilon). \tag{3.73}$$

The variables g and p are defined in (3.11) and in (2.3) respectively as

$$\frac{(1 - b_1 x)}{(1 - a_2 x)} g = \frac{y(e^{-\epsilon} x)}{y(x)} \quad \text{for} \quad x = f^{-1}, \quad \text{and} \quad p = \frac{y'(x)}{y(x)} \quad \text{for} \quad x = q, \tag{3.74}$$

hence we have the relation

$$\frac{f - b_1}{f - a_2} g = 1 - \epsilon q p + O(\epsilon^2). \tag{3.75}$$

Through (3.72), (3.73) and (3.75), we obtain the P_{VI} equation given by the Hamiltonian in (3.76)

$$H = \frac{q(q-1)(q-t)}{t(t-1)} \left\{ p^2 \left(\frac{m+n+1}{q} + \frac{\alpha}{q-1} + \frac{\beta-1}{q-t} \right) p + \frac{m(\alpha+\beta+n)}{q(q-1)} \right\}. \tag{3.76}$$

3.4 Relation to 2 × 2 Lax Form

Originally [52] the *q-Garnier system* was formulated in 2×2 matrix Lax pair:

$$Y(qx) = \mathcal{A}(x)Y(x), \quad \overline{Y}(x) = \mathcal{B}(x)Y(x). \tag{3.77}$$

Where the matrix $\mathcal{A}(x)$ is characterized by the following conditions:[4]

$$\mathcal{A}(x) = \mathcal{A}_0 + \mathcal{A}_1 x + \cdots + \mathcal{A}_{N+1}x^{N+1},$$
$$\mathcal{A}_{N+1} = \text{diag}(\kappa_1, \kappa_2), \quad \mathcal{A}_0 \text{ has the eigenvalues } \theta_1, \theta_2, \tag{3.78}$$
$$\det\mathcal{A}(x) = \kappa_1\kappa_2 \prod_{i=1}^{2N+2}(x-\alpha_i).$$

The matrix $\mathcal{B}(x)$ should be specified suitably depending the direction of the deformation $\overline{*}$. The compatibility of these equations is given by

$$\overline{\mathcal{A}}(x)\mathcal{B}(x) = \mathcal{B}(qx)\mathcal{A}(x). \tag{3.79}$$

Proposition 3.3 *For the first component $y_1(x)$ of $Y(x) = \begin{bmatrix} y_1(x) \\ y_2(x) \end{bmatrix}$, we have*

$$L_1 : F(\tfrac{x}{q})y_1(qx) - \{a_{11}(x)a_{12}(\tfrac{x}{q}) + a_{12}(x)a_{22}(\tfrac{x}{q})\}y_1(x) + F(x)|\mathcal{A}(\tfrac{x}{q})|y_1(\tfrac{x}{q}) = 0,$$
$$L_2 : G(x)y_1(x) - b_{12}(x)y_1(qx) + F(x)\overline{y}_1(x) = 0,$$
$$L_3 : \overline{F}(x)|\mathcal{B}(x)|y_1(qx) - b_{12}(qx)|\mathcal{A}(x)|\overline{y}_1(x) + G(x)\overline{y}_1(qx) = 0,$$
$$\tag{3.80}$$

where $\mathcal{A} = (a_{ij})$, $\mathcal{B} = (b_{ij})$, and

$$F(x) = a_{12}(x), \quad G(x) = a_{11}(x)b_{12}(x) - a_{12}(x)b_{11}(x). \tag{3.81}$$

Proof In the proof we use the notation $f^* = f\big|_{x \to qx}$, $f_x = f\big|_{x \to x/q}$. In components, the equations (3.77) are written as

---

[4] In view of the symmetry between $x = 0$ and $x = \infty$, it is convenient to put $\mathcal{A}_0$ [$\mathcal{A}_{N+1}$, resp.] into upper [lower, resp.] triangular form by a gauge transformation.

$$A_1 \; : \; y_1^x = a_{11}y_1 + a_{12}y_2, \tag{3.82}$$
$$A_2 \; : \; y_2^x = a_{21}y_1 + a_{22}y_2,$$
$$B_1 \; : \; \overline{y_1} = b_{11}y_1 + b_{12}y_2,$$
$$B_2 \; : \; \overline{y_2} = b_{21}y_1 + b_{22}y_2.$$

From A_1 we have $y_2 = \frac{1}{a_{12}}(y_1^x - a_{11}y_1)$. Substituting this into A_2, B_1, we have

$$L_1^x : \quad Fy_1^{xx} - (a_{12}a_{11}^x + a_{12}^x a_{22})y_1^x + F^x|\mathcal{A}|y_1 = 0, \tag{3.83}$$
$$L_2 : \quad Gy_1 - b_{12}y_1^x + F\overline{y_1} = 0.$$

Eliminating y_1^{xx} from L_1^x, L_2^x, we have

$$F\overline{y_1}^x - (a_{12}b_{11}^x + a_{22}b_{12}^x)y_1^x + b_{12}^x|\mathcal{A}|y_1 = 0, \tag{3.84}$$

and eliminating y_1 from this equation and L_2, we obtain

$$L_3 : \quad \overline{F}|\mathcal{B}|y_1^x - b_{12}^x|\mathcal{A}|\overline{y_1} + G\overline{y_1}^x = 0, \tag{3.85}$$

where $\overline{F} = (\mathcal{B}^x \mathcal{A}\mathcal{B}^{-1})_{12}$ due to the compatibility (3.79). $\qquad\square$

Proposition 3.4 *From the compatibility condition* (3.79) *we have*

$$F(x)\overline{F}(x)|\mathcal{B}(x)| - b_{12}(x)b_{12}(qx)|\mathcal{A}(x)| = 0 \quad for \; G(x) = 0, \tag{3.86}$$

$$G(x)\overline{G}(x) - b_{12}(qx)\overline{b}_{12}(x)|\mathcal{A}(x)| = 0 \quad for \; \overline{F}(x) = 0. \tag{3.87}$$

Though the relations (3.86), (3.87) can be easily obtained from the compatibility conditions of scalar equations L_2, L_3. We give another proof of them based on the matrix compatibility (3.79). This proof gives not only the divisibility but also the explicit quotients.

Proof For 2×2 matrices $M_1, \ldots, M_j$, define m_{ij} ($1 \leq i, j \leq 4$) as

$$m_{ij} = (M_i M_j^{-1})_{12}|M_j| = -(M_i)_{11}(M_j)_{12} + (M_i)_{12}(M_j)_{11} = -m_{ji}. \tag{3.88}$$

These are minor determinants of the 2×4 matrix

$$\begin{bmatrix} (M_1)_{12} & (M_2)_{12} & (M_3)_{12} & (M_4)_{12} \\ (M_1)_{11} & (M_2)_{11} & (M_3)_{11} & (M_4)_{11} \end{bmatrix}.$$

Then, by the Plücker relation, we have

$$m_{12}m_{34} - m_{13}m_{24} + m_{14}m_{23} = 0. \tag{3.89}$$

Putting $M_1 = \mathcal{A}$, $M_2 = \mathcal{B}$, $M_3 = \mathcal{B}^x\mathcal{A}$ and $M_4 = I$, we have

$$m_{12} = -m_{21} = (\mathcal{A}\mathcal{B}^{-1})_{12}|\mathcal{B}| = -G, \qquad (3.90)$$

$$m_{34} = -m_{43} = (\mathcal{B}^x\mathcal{A})_{12},$$

$$m_{13} = -m_{31} = -b_{12}^x|\mathcal{A}|,$$

$$m_{24} = -m_{42} = b_{12},$$

$$m_{14} = -m_{41} = (\mathcal{A})_{12} = F,$$

$$m_{23} = -m_{32} = -(\overline{\mathcal{A}})_{12}|\mathcal{B}| = -\overline{F}|\mathcal{B}|,$$

and

$$G(\mathcal{B}^x\mathcal{A})_{12} - b_{12}b_{12}^x|\mathcal{A}| + F\overline{F}|\mathcal{B}| = 0. \qquad (3.91)$$

Hence we obtain (3.86).

Similarly, putting $M_1 = \mathcal{A}$, $M_2 = \mathcal{B}$, $M_3 = \mathcal{B}^x\mathcal{A}$ and $M_4 = \overline{\mathcal{B}}\mathcal{B}$, we have

$$m_{12} = -m_{21} = (\mathcal{A}\mathcal{B}^{-1})_{12}|\mathcal{B}| = -G, \qquad (3.92)$$

$$m_{34} = -m_{43} = (\overline{\mathcal{A}\mathcal{B}}^{-1})_{12}|\overline{\mathcal{B}}||\mathcal{B}| = -\overline{G}|\mathcal{B}|,$$

$$m_{13} = -m_{31} = -b_{12}^x|\mathcal{A}|,$$

$$m_{24} = -m_{42} = -(\overline{\mathcal{B}})_{12}|\mathcal{B}|,$$

$$m_{14} = -m_{41} = -(\overline{\mathcal{B}}\mathcal{B}\mathcal{A}^{-1})_{12}|\mathcal{A}|,$$

$$m_{23} = -m_{32} = -\overline{F}|\mathcal{B}|,$$

and

$$G\overline{G} - b_{12}b_{12}^x|\mathcal{A}| + (\overline{\mathcal{B}}\mathcal{B}\mathcal{A}^{-1})_{12}|\mathcal{A}|\overline{F} = 0. \qquad (3.93)$$

Hence we obtain (3.87). $\qquad\qquad\qquad\qquad\qquad\qquad\qquad\qquad\qquad\qquad\qquad$ □

Chapter 4
Padé Interpolation

Abstract The Padé approximation has a discrete analog: the Padé interpolation. In fact, the latter was known before the former. We study the discrete Painlevé equations using the Padé interpolation. In this chapter, we consider the interpolation on a q-grid defined by $x_s = q^s$ $(s = 0, 1, 2, \ldots)$.

4.1 Cauchy–Jacobi Formula

Consider the following interpolation problem by rational functions.

> For a given sequence ψ_s, we consider polynomials $P(x)$ and $Q(x)$ of degree m and n, by the following interpolation condition:
>
> $$\psi_s = \frac{P(x_s)}{Q(x_s)} \quad (s = 0, 1, \ldots, m+n), \tag{4.1}$$
>
> where x_s are given points (grid) such that $x_s \neq x_{s'}$ for $s \neq s'$.

This is a generic case of the *Padé interpolation* and its solution is classically known as the *Cauchy–Jacobi formula* due to Cauchy [3] and Jacobi [18].

Proposition 4.1 *[3] The polynomials $P(x)$, $Q(x)$ are given as follows:*

$$P(x) = \sum_{J:|J|=n+1} \Delta_J^2 \prod_{s \in J} u_s \prod_{s \notin J} (x - x_s),$$
$$Q(x) = \sum_{I:|I|=n} \Delta_I^2 \prod_{s \in I} u_s \prod_{s \in I} (x - x_s), \tag{4.2}$$

where $u_s = \frac{\psi_s}{f'(x_s)}$ and $f(x) = \prod_{i=0}^{m+n}(x - x_i)$. I, J are subsets of $\{0, 1, \ldots, m+n\}$ and $\Delta_I = \prod_{i,j \in I, i<j}(x_i - x_j)$.

H. Nagao and Y. Yamada, *Padé Methods for Painlevé Equations*,
SpringerBriefs in Mathematical Physics,
https://doi.org/10.1007/978-981-16-2998-3_4

In the case of $n = 0$, $Q = 1$, the formula reduces to the well-known Lagrange formula

$$P(x) = \sum_{s=0}^{m} \psi_s \prod_{t(\neq s)=0}^{m} \frac{x - x_t}{x_s - x_t}. \tag{4.3}$$

Proof To check Cauchy's formula, we compute $P(x_i)$. In the sum over J, the nonzero terms are such that $i \in J$, hence we can put $J = I + \{i\}$ (disjoint union), and obtain

$$P(x_i) = \sum_{i \notin I, |I|=n} \Delta_I^2 \prod_{s \in I} (x_i - x_s)^2 u_i \prod_{s \in I} u_s \prod_{s \notin I, s \neq i} (x_i - x_s). \tag{4.4}$$

Then noting that $f'(x_i) = \prod_{s \neq i} (x_i - x_s) = \prod_{s \in I} (x_i - x_s) \prod_{s \notin I, s \neq i} (x_i - x_s)$, we have the desired result:

$$P(x_i) = u_i f'(x_i) \sum_{i \notin I, |I|=n} \Delta_I^2 \prod_{s \in I} u_s \prod_{s \in I} (x_i - x_s) = \psi_i Q(x_i). \tag{4.5}$$

Later, Jacobi gave a determinant expression of Cauchy's formula.

Proposition 4.2 *[18] Polynomials* $P(x)$, $Q(x)$ *have the following determinant expression:*

$$P(x) = f(x) \det \left[\sum_{s=0}^{m+n} u_s \frac{x_s^{i+j}}{x - x_s} \right]_{i,j=0}^{n}, \quad Q(x) = \det \left[\sum_{s=0}^{m+n} u_s x_s^{i+j} (x - x_s) \right]_{i,j=0}^{n-1}. \tag{4.6}$$

Proof To see this, we note that the matrix in the expression for $P(x)$ is a product of three matrices $V_1 D V_2$ such that:

- V_1: $(n + 1) \times (m + n + 1)$ matrix with (i, s) element x_s^i,
- D: $(m + n + 1)$ diagonal matrix with (s, s) element $u_s/(x - x_s)$,
- V_2: $(m + n + 1) \times (n + 1)$ matrix with (s, j) element x_s^j.

Then we have

$$\det(V_1 D V_2) = \sum_{J, |J|=n+1} \Delta_J \left(\prod_{s \in J} \frac{u_s}{x - x_s} \right) \Delta_J = \frac{P(x)}{f(x)}. \tag{4.7}$$

The determinant expression for the $Q(x)$ polynomial can be checked similarly. $\quad\square$

In the following, we use the formula in the case where the interpolation points are given by the q-grid: $x_s = q^s$. Then, using the formulas

$$f(x) = \prod_{s=0}^{m+n} (x - q^s), \quad f'(x_s) = \frac{(q)_s (q)_{m+n}}{q^s (q^{-(m+n)})_s}, \tag{4.8}$$

we have

$$
P(x) = \frac{f(x)}{(q)_{m+n}^{n+1}} \det \left[\sum_{s=0}^{m+n} \psi_s \frac{(q^{-(m+n)})_s}{(q)_s} \frac{q^{s(i+j+1)}}{x - q^s} \right]_{i,j=0}^{n},
$$

$$
Q(x) = \frac{1}{(q)_{m+n}^{n}} \det \left[\sum_{s=0}^{m+n} \psi_s \frac{(q^{-(m+n)})_s}{(q)_s} q^{s(i+j+1)}(x - q^s) \right]_{i,j=0}^{n-1}.
$$

(4.9)

Remark 4.1 Instead of the monomial x_s^i, any monic polynomial of x_s of degree i can be used as the (i, s) element of the Vandermonde matrix V_1 (and similar for V_2).

4.2 Application to q-Garnier System

In this chapter, we consider the following Padé interpolation problem.

For complex parameters a_i, b_i, c, we put

$$
\psi(x) = c^{\log_q x} \prod_{i=1}^{N} \frac{(a_i x, b_i)_\infty}{(a_i, b_i x)_\infty}, \quad \psi_s = \psi(q^s) = c^s \prod_{i=1}^{N} \frac{(b_i)_s}{(a_i)_s}. \tag{4.10}
$$

Define polynomials $P(x)$ and $Q(x)$ of degree m and n by the following Padé interpolation condition:

$$
\psi(x_s) = \frac{P(x_s)}{Q(x_s)}, \quad x_s = q^s \quad (s = 0, 1, \dots m + n). \tag{4.11}
$$

The common normalization of the polynomials $P(x)$ and $Q(x)$ is fixed as $P(0) = 1$.

As before, the shift T is given by (3.9).

Proposition 4.3 *For $y(x) = P(x)$ and $y(x) = \psi(x)Q(x)$, we have the following linear relations:*

$$
L_2 : (g_0)_1 F(x)\overline{y}(x) - (\frac{x}{q^{m+n}})_1 A_1(x) y(qx) + (b_1 x)_1 G(x) y(x) = 0, \tag{4.12}
$$

$$
L_3 : (\frac{g_0}{c})_1 \overline{F}(\frac{x}{q}) y(x) + \frac{1}{c}(a_1 x)_1 G(\frac{x}{q})\overline{y}(x) - (x)_1 B_1(\frac{x}{q})\overline{y}(\frac{x}{q}) = 0,
$$

where

$$A(x) = \prod_{j=1}^{N}(a_j x)_1, \quad B(x) = \prod_{j=1}^{N}(b_j x)_1, \quad F(x) = 1 + \sum_{j=1}^{N} f_j x^j, \quad (4.13)$$

$$A_i(x) = \frac{A(x)}{(a_i x)_1}, \quad B_i(x) = \frac{B(x)}{(b_i x)_1}, \quad G(x) = \sum_{j=0}^{N-1} g_j x^j. \quad (4.14)$$

Here $f_1, \ldots, f_N, g_0, \ldots, g_{N-1}$ are some constants.

Proof The derivations are similar to that in previous section. The only difference is the divisibility by x^{m+n} for instance will be changed to the divisibility by $\prod_{s=0}^{m+n}(1 - \frac{x}{q^s})$. In the proof, we use the up/down-shift notation $f^* = f|_{x \to qx}$ and $f_x = f|_{x \to x/q}$. As before, L_2 and L_3^x are given by

$$L_2 : D_1 \bar{y} - D_2 y^* + D_3 y = 0, \quad (4.15)$$
$$L_3^x : \overline{D_1} y^* + D_3 \bar{y}^* - D_2^x \bar{y} = 0, \quad (4.16)$$

where

$$D_1 = |\mathbf{u}, \mathbf{u}^*|, \quad D_2 = |\mathbf{u}, \bar{\mathbf{u}}|, \quad D_3 = |\mathbf{u}^*, \bar{\mathbf{u}}|, \quad \mathbf{u} = \begin{bmatrix} P \\ \psi Q \end{bmatrix}. \quad (4.17)$$

The Casorati determinants $D_i(x)$ can be computed by using the condition (4.11) and the relations

$$\frac{\psi^x}{\psi} = c\frac{B}{A}, \quad \frac{\bar{\psi}}{\psi} = \frac{(a_1, b_1 x)_1}{(a_1 x, b_1)_1}. \quad (4.18)$$

The results are given as follows:

$$D_1 = \frac{\psi}{A}\{cBPQ^* - AP^*Q\} = w_0 \frac{\psi}{A} \prod_{i=0}^{m+n-1}(\frac{x}{q^i})_1 F, \quad (4.19)$$

$$D_2 = \frac{\psi}{(a_1 x, b_1)_1}\{(a_1, b_1 x)_1 P\bar{Q} - (a_1 x, b_1)_1 \bar{P}Q\} = w_1 \frac{\psi}{(a_1 x)_1}\prod_{i=0}^{m+n}(\frac{x}{q^i})_1,$$

$$D_3 = \frac{\psi}{(b_1)_1 A}\{(a_1, b_1 x)_1 A_1 P^x \bar{Q} - c(b_1)_1 B\bar{P}Q^*\}$$

$$= w_1 \frac{\psi}{A}(b_1 x)_1 \prod_{i=0}^{m+n-1}(\frac{x}{q^i})_1 G,$$

where w_0, w_1 are some constants in x. The ratios of these constants w_0/w_1, $\bar{w}_0/w_1$ can be determined as (4.12) using a solution $y(x) = P(x) = 1 + O(x)$. □

The equations L_2, L_3 in (4.12) can be regarded as a Lax pair for the q-Garnier system, since they are the same form as (3.11), (3.12) under the identification of the parameters:

	$\{a_i\}$	$\{b_i\}$	$\{c_i\}$	$\{d_i\}$
Chap. 3	$a_1, \ldots, a_N, a_{N+1}$	$b_1, \ldots, b_N, b_{N+1}$	$q^m, q^n \prod_{i=1}^{N+1} \frac{b_i}{a_i}$	$1, q^{m+n+1}$
Chap. 4	$a_1, \ldots, a_N, q^{-m-n}$	$b_1, \ldots, b_N, q$	$q^m, q^n c \prod_{i=1}^{N} \frac{b_i}{a_i}$	$1, c$

$$(4.20)$$

Here, the parameters $\{c_i\}$ and $\{d_i\}$ represent the *multiplicative exponents* (the eigenvalues of the operator $T_x : x \mapsto qx$) at $x \to \infty$ and $x \to 0$ respectively. Accordingly, the following compatibility conditions can also be regarded as a q-Garnier system.

Proposition 4.4 *The coefficients $f_1, \ldots, f_N$ and $g_0, \ldots, g_{N-1}$ of the polynomials $F(x), G(x)$ satisfy the following relations:*

$$(qx, \frac{x}{q^{m+n}})_1 A_1(x) B_1(x) - \frac{1}{c}(a_1 x, b_1 x)_1 G(x)\underline{G}(x) = 0 \quad \text{for} \quad F(x) = 0, \quad (4.21)$$

$$(qx, \frac{x}{q^{m+n}})_1 A_1(x) B_1(x) - (g_0, \frac{g_0}{c})_1 F(x)\overline{F}(x) = 0 \quad \text{for} \quad G(x) = 0, \quad (4.22)$$

$$(g_0, \frac{g_0}{c})_1 f_N \overline{f}_N = \frac{q a_1 b_1}{c}\left(g_{N-1} - \frac{c \prod_{i=2}^{N}(-b_i)}{a_1 q^m}\right)\left(g_{N-1} - \frac{\prod_{i=2}^{N}(-a_i)}{b_1 q^n}\right). \quad (4.23)$$

Proof This follows from the compatibility of the equations L_2, L_3 in (4.12). ☐

Example

The system of equations given above is the q-Garnier system in $2N$-variables, and as discussed in the previous section, the $N = 1$ case is the q-P_{VI} and the $N = 2$ case has a reduction to q-$P(E_6^{(1)})$. Here we will consider a reduction from the $N = 3$ case to a 2-variable system according to [38, 40, 41].

We put two constraints on parameters $c = 1$ and $q^m a_1 a_2 a_3 = q^n b_1 b_2 b_3$ to make the exponents at $x = 0$ and $x = \infty$ both degenerate. Then the polynomial $F(x), G(x)$ can be reduced to $F(x) = w(1 - fx)$ and $G(x) = 1 + gx + \kappa x^2$. ($\kappa = \frac{b_2 b_3}{q^m a_1}$) ($w$ is a gauge freedom). For the deformation direction $T = T_{a_1} T_{b_1}$ we have the contiguity relations

$$L_2 : wx(1 - fx)\overline{y}(x) - \prod_{i=2}^{4}(1 - a_i x)y(qz) + (1 - b_1 z)G(x)y(x) = 0,$$

$$L_3 : \overline{w}\frac{x}{q}(1 - \overline{f}\frac{x}{q})y(x) + (1 - a_1 x)G(\frac{z}{q})\overline{y}(x) - \prod_{i=2}^{4}(1 - b_i\frac{x}{q})\overline{y}(\frac{x}{q}) = 0,$$

$$(4.24)$$

where $a_4 = q^{-m-n}$, $b_4 = q$. As the compatibility condition, the evolution equation is given by

$$(f^2 + gf + \kappa)(f^2 + \underline{g}f + q\kappa) = \frac{\prod_{i=2}^{4}(f - a_i)(f - b_i)}{(f - a_1)(f - b_1)},$$

(4.25)

$$\frac{x_1^2(1 - fx_1)(1 - \overline{f}x_1)}{x_2^2(1 - fx_2)(1 - \overline{f}x_2)} = \prod_{i=2}^{4} \frac{(x_1 - a_i)(x_1 - b_i)}{(x_2 - a_i)(x_2 - b_i)},$$

where x_1, x_2 are solutions of $G(x) = 0$. The configuration of singular points is given by two points on a line $g = \infty$ and six points on a curve $f^2 + gf + \kappa = 0$.

In order to see the relation to the q-Painlevé equation, we consider another deformation direction $T' = T_{a_1} T_{a_2} T_{b_1} T_{b_2}$ ($\hat{x} = T'(x)$). Then the new Lax pair is

$$L_2' : ux(1 - fx)\hat{y}(z) - (\xi - x) \prod_{i=3,4}(1 - a_i x)y(qz)$$

$$+ (\xi - \eta x) \prod_{i=1,2}(1 - b_i x)y(x) = 0,$$

$$L_3' : \hat{u}\frac{x}{q}(1 - \hat{f}\frac{x}{q})y(z) - (\xi - \eta\frac{x}{q}) \prod_{i=1,2}(1 - a_i x)\hat{y}(z)$$

$$+ (\xi - x) \prod_{i=3,4}(1 - b_i \frac{x}{q})\hat{y}(\frac{x}{q}) = 0,$$

(4.26)

where $\eta = \frac{c_3}{a_1 a_2 q^m}$, ξ is a variable and u is a gauge freedom. The evolution equation is given by

$$\frac{(f\xi - \eta)(fT'^{-1}(\xi) - q^2\eta)}{(f\xi - 1)(fT'^{-1}(\xi) - q)} = \frac{\prod_{i=3,4}(f - a_i)(f - b_i)}{\prod_{i=1,2}(f - a_i)(f - b_i)},$$

(4.27)

$$\frac{(f\xi - \eta)(\hat{f}\xi - \eta)}{(f\xi - 1)(\hat{f}\xi - 1)} = \frac{\prod_{i=3,4}(1 - \xi a_i)(1 - \xi b_i)}{\prod_{i=1,2}(1 - \xi a_i/\eta)(1 - \xi b_i/\eta)}.$$

This equation is the standard form of the q-Painlevé equation of type $E_7^{(1)}$ in [12, 25, 51, 64], and the configuration of singular points is given by two curves $f\xi = 1$ and $f\xi = \eta$. Comparing the L_2 and L_2' at $x = 1/f$, the relation between the variables g, ξ is given by

$$\frac{(f - a_2)(f - b_2)}{f^2 + gf + \kappa} = \frac{f\xi - 1}{f\xi - \eta}.$$

(4.28)

As a result, the Eq. (4.25) is a variation of the q-Painlevé equation of type $E_7^{(1)}$ whose deformation direction is different to the standard one.

We will derive the explicit forms of variables $\{f_i, g_i\}$ appearing in the Casorati determinants D_1 and D_3 (4.19). Then, these formulas give a special solution for the q-Garnier system. This solution corresponds to a different specialization compared to the solution given in Chap. 3.

4.3 Special Solutions

Lemma 4.1 *We note that the polynomials $P(x)$ and $Q(x)$ have the following special values:*

$$P(\frac{1}{a_i}) = \frac{(a_i)_{m+n+1}}{a_i^m (a_i)_1^{n+1}(q)_{m+n}^{n+1}} T_{a_i}(\tau_{m,n}), \quad Q(\frac{q}{a_i}) = \frac{q^n (\frac{a_i}{q})_1^n}{a_i^n (q)_{m+n}^n} T_{a_i}^{-1}(\tau_{m+1,n-1}),$$

$$\text{(4.29)}$$

$$P(\frac{q}{b_i}) = \frac{q^m (\frac{b_i}{q})_{m+n+1}}{b_i^m (\frac{b_i}{q})_1^{n+1}(q)_{m+n}^{n+1}} T_{b_i}^{-1}(\tau_{m,n}), \quad Q(\frac{1}{b_i}) = \frac{(b_i)_1^n}{b_i^n (q)_{m+n}^n} T_{b_i}(\tau_{m+1,n-1}),$$

for $i = 1, \ldots, N$, where

$$\tau_{m,n} = \det\left[{}_{N+1}\varphi_N\left(\begin{matrix} b_1, \ldots, b_N, q^{-(m+n)} \\ a_1, \ldots, a_N \end{matrix}, cq^{i+j+1} \right) \right]_{i,j=0}^n, \quad \text{(4.30)}$$

and ${}_{N+1}\varphi_N$ is the generalized q-hypergeometric series *[10]*

$$_{N+1}\varphi_N\left(\begin{matrix} \alpha_1, \ldots, \alpha_{N+1} \\ \beta_1, \ldots, \beta_N \end{matrix}, x \right) = \sum_{s=0}^{\infty} \frac{(\alpha_1, \ldots, \alpha_{N+1})_s}{(\beta_1, \ldots, \beta_N, q)_s} x^s. \quad \text{(4.31)}$$

Proof Applying $\psi_s = c^s \prod_{i=1}^N \frac{(b_i)_s}{(a_i)_s}$ to the formula (4.9), we obtain the determinants $\tau_{m,n}$ except for a factor $(x - q^s)^{\pm 1}$. For the special values $x = \frac{1}{a_i}, \frac{1}{b_i}, \frac{q}{a_i}, \frac{q}{b_i}$, this extra factor can be absorbed by a suitable shift of the parameter. For instance, using $T_{a_i}(\psi_s) = \frac{1-a_i}{1-a_i q^s}\psi_s$, we obtain the expression for $P(\frac{1}{a_i})$. Other cases are similar. □

Remark 4.2 Different but closely related q-isomonodromic equations which also have special solutions written as ${}_{N+1}\varphi_N$ were studied by T. Suzuki (see [56] for example).

Lemma 4.2

$$\frac{F(\frac{1}{a_i})}{F(\frac{1}{b_j})} = -c \prod_{s=0}^{m+n-1} \frac{(\frac{1}{b_j q^s})_1}{(\frac{1}{a_i q^s})_1} \frac{B(\frac{1}{a_i})}{A(\frac{1}{b_j})} \frac{P(\frac{1}{a_i})Q(\frac{q}{a_i})}{P(\frac{q}{b_j})Q(\frac{1}{b_j})} \quad (i, j = 1, \ldots, N), \quad \text{(4.32)}$$

$$G(\frac{1}{a_i}) = -\frac{c(b_1)_1 B_1(\frac{1}{a_i})}{(a_1, \frac{b_1}{a_1})_1} \frac{\prod_{s=0}^{m+n} (\frac{1}{a_1 q^s})_1}{\prod_{s=0}^{m+n-1} (\frac{1}{a_i q^s})_1} \frac{\overline{P}(\frac{1}{a_i})Q(\frac{q}{a_i})}{P(\frac{1}{a_1})\overline{Q}(\frac{1}{a_1})} \quad (i = 2, \ldots, N), \quad \text{(4.33)}$$

$$G(\frac{1}{b_i}) = -\frac{\prod_{s=0}^{m+n} (\frac{1}{b_1 q^s})_1}{\prod_{s=0}^{m+n-1} (\frac{1}{b_i q^s})_1} \frac{(a_1)_1 A_1(\frac{1}{b_i})}{(\frac{a_1}{b_1}, b_1)_1} \frac{P(\frac{q}{b_i})\overline{Q}(\frac{1}{b_i})}{\overline{P}(\frac{1}{b_1})Q(\frac{1}{b_1})} \quad (i = 2, \ldots, N). \quad \text{(4.34)}$$

Proof From (4.19), for the polynomials $F(x)$, $G(x)$ we have the lemma. □

Lemma 4.3 *We have the following expressions for the special values of polynomials* $F(x)$, $G(x)$:

$$\frac{F(\frac{1}{a_i})}{F(\frac{1}{b_j})} = \alpha \frac{T_{a_i}(\tau_{m,n})T_{a_i}^{-1}(\tau_{m+1,n-1})}{T_{b_j}^{-1}(\tau_{m,n})T_{b_j}(\tau_{m+1,n-1})} \quad (i,j = 1,\ldots,N), \tag{4.35}$$

$$G(\frac{1}{a_i}) = \beta \frac{T_{a_i}(\overline{\tau}_{m,n})T_{a_i}^{-1}(\tau_{m+1,n-1})}{T_{a_1}(\tau_{m,n})T_{a_1}^{-1}(\overline{\tau}_{m+1,n-1})} \quad (i = 2,\ldots,N), \tag{4.36}$$

$$G(\frac{1}{b_i}) = \gamma \frac{T_{b_i}^{-1}(\tau_{m,n})T_{b_i}(\overline{\tau}_{m+1,n-1})}{T_{b_1}^{-1}(\overline{\tau}_{m,n})T_{b_1}(\tau_{m+1,n-1})} \quad (i = 2,\ldots,N), \tag{4.37}$$

where

$$\alpha = -cq^{n-m}\frac{(a_iq^{m+n})_1(\frac{b_j}{q})_1^n(\frac{a_i}{q})_1^n}{(a_i)_1^{n+1}(b_j)_1^n}\frac{B(\frac{1}{a_i})}{A(\frac{1}{b_j})},$$

$$\beta = c\frac{(b_1,a_iq^{m+n})_1(\frac{a_i}{q})_1^n B_1(\frac{1}{a_i})}{a_1q^m(\frac{b_1}{a_1})_1(a_i)_1^{n+1}}, \quad \gamma = \frac{(a_1)_1(b_i)_1^n A_1(\frac{1}{b_i})}{b_1q^n(\frac{a_1}{b_1})_1(\frac{b_i}{q})_1^n}. \tag{4.38}$$

Proof Substituting the special values (4.29) into the expressions (4.32)–(4.34) respectively. $\qquad\square$

Remark 4.3 Using the following relation between the q-Appell–Lauricella function φ_D and the generalized q-hypergeometric function $_{N+1}\varphi_N$,

$$_{N+1}\varphi_N\left(\begin{matrix} a,b_1\ldots,b_N \\ c_1,\ldots,c_N \end{matrix}, u\right) = \frac{(au)_\infty}{(u)_\infty}\prod_{k=1}^N\frac{(b_k)_\infty}{(c_k)_\infty}\varphi_D(u,\frac{c_1}{b_1},\ldots,\frac{c_N}{b_N}, au; b_1,\ldots,b_N), \tag{4.39}$$

the special solution obtained above can be rewritten as

$$\tau_{m,n} = \left(\frac{(cq^{-m-n})_\infty}{(q)_\infty}\prod_{i=1}^N\frac{(b_i)_\infty}{(a_i)_\infty}\right)^{n+1}\tilde{\tau}_{m,n}, \quad \tilde{\tau}_{m,n} = \det\left[h_{i+j}\right]_{i,j=0}^n,$$

$$h_k = \frac{(cq)_k}{(cq^{-m-n+1})_k}\varphi_D(cq^{k+1},\frac{a_1}{b_1},\ldots,\frac{a_N}{b_N},cq^{k+1-m-n};b_1,\ldots,b_N), \tag{4.40}$$

$$\frac{F(\frac{1}{a_i})}{F(\frac{1}{b_j})} = \tilde{\alpha}\frac{T_{a_i}(\tilde{\tau}_{m,n})T_{a_i}^{-1}(\tilde{\tau}_{m+1,n-1})}{T_{b_j}^{-1}(\tilde{\tau}_{m,n})T_{b_j}(\tilde{\tau}_{m+1,n-1})} \quad (i,j = 1,\ldots,N), \tag{4.41}$$

$$G(\frac{1}{a_i}) = \tilde{\beta}\frac{T_{a_i}(\overline{\tilde{\tau}}_{m,n})T_{a_i}^{-1}(\tilde{\tau}_{m+1,n-1})}{T_{a_1}(\tilde{\tau}_{m,n})T_{a_1}^{-1}(\overline{\tilde{\tau}}_{m+1,n-1})} \quad (i = 2,\ldots,N), \tag{4.42}$$

$$G(\frac{1}{b_i}) = \tilde{\gamma} \frac{T_{b_i}^{-1}(\tilde{\tau}_{m,n}) T_{b_i}(\tilde{\tilde{\tau}}_{m+1,n-1})}{T_{b_1}^{-1}(\tilde{\tilde{\tau}}_{m,n}) T_{b_1}(\tilde{\tau}_{m+1,n-1})} \quad (i = 2, \ldots, N), \qquad (4.43)$$

and

$$\tilde{\alpha} = -cq^{n-m} \frac{(a_i q^{m+n})_1}{(\frac{b_j}{q})_1} \frac{B(\frac{1}{a_i})}{A(\frac{1}{b_j})}, \quad \tilde{\beta} = c \frac{(a_i q^{m+n})_1 B_1(\frac{1}{a_i})}{a_1 q^m (\frac{b_1}{a_1})_1}, \quad \tilde{\gamma} = \frac{(\frac{b_i}{q})_1 A_1(\frac{1}{b_i})}{b_1 q^n (\frac{a_1}{b_1})_1}.$$
$$(4.44)$$

We give a proof of the relation (4.39) in a more general setting in the next section.

4.4 A Duality Between q-Appell–Lauricella and q-HG Series

In this section the index i (resp. j) always runs over $1 \leq i \leq M$ (resp. $1 \leq j \leq N$), and $\{a_1, \ldots, a_M\}$ (resp. $\{b_1, \ldots, b_N\}$) will simply be written as $\{a_i\}$ (resp. $\{b_j\}$).

Following [49], we define a series $\mathcal{F}_{M,N}$ as

$$\mathcal{F}_{M,N}\left(\begin{matrix}\{a_i\}, \{b_j\}\\ \{c_i\}\end{matrix}; \{y_j\}\right) = \sum_{n_j \geq 0} \prod_{i=1}^{M} \frac{(a_i)_{|n|}}{(c_i)_{|n|}} \prod_{j=1}^{N} \frac{(b_j)_{n_j}}{(q)_{n_j}} \prod_{j=1}^{N} y_j^{n_j}, \qquad (4.45)$$

where $|n| = \sum_{j=1}^{N} n_j$. The series $\mathcal{F}_{M,N}$ is a common generalization of q-Appell–Lauricella function φ_D and generalized q-hypergeometric function $_{M+1}\varphi_M$, i.e.

$$\mathcal{F}_{1,N}\left(\begin{matrix}a, \{b_j\}\\ c\end{matrix}; \{y_j\}\right) = \varphi_D\left(\begin{matrix}a, \{b_j\}\\ c\end{matrix}; \{y_j\}\right) = \sum_{n_j \geq 0} \frac{(a)_{|n|}}{(c)_{|n|}} \prod_{j=1}^{N} \frac{(b_j)_{n_j}}{(q)_{n_j}} \prod_{j=1}^{N} y_j^{n_j}, \quad (4.46)$$

$$\mathcal{F}_{M,1}\left(\begin{matrix}\{a_i\}, b\\ \{c_i\}\end{matrix}; y\right) = {}_{M+1}\varphi_M\left(\begin{matrix}\{a_i\}, b\\ \{c_i\}\end{matrix}; y\right) = \sum_{n \geq 0} \prod_{i=1}^{M} \frac{(a_i)_n}{(c_i)_n} \frac{(b)_n}{(q)_n} y^n. \qquad (4.47)$$

Proposition 4.5 *[49] The following duality relation holds:*

$$\mathcal{F}_{N,M}\left(\begin{matrix}\{y_j\}, \{a_i\}\\ \{b_j y_j\}\end{matrix}; \{x_i\}\right) = \frac{(\{y_j\}, \{a_i x_i\})_\infty}{(\{b_j y_j\}, \{x_i\})_\infty} \mathcal{F}_{M,N}\left(\begin{matrix}\{x_i\}, \{b_j\}\\ \{a_i x_i\}\end{matrix}; \{y_j\}\right). \qquad (4.48)$$

Proof Consider the sum:

$$S = \sum_{m_i} \sum_{n_j} \prod_{i=1}^{M} \frac{(a_i)_{m_i}}{(q)_{m_i}} x_i^{m_i} \prod_{j=1}^{N} \frac{(b_j)_{n_j}}{(q)_{n_j}} y_j^{n_j} q^{|m||n|}, \qquad (4.49)$$

where $|m| = \sum_{i=1}^{M} m_i$. Using the q-binomial theorem

$$\sum_{n=0}^{\infty} \frac{(b)_n}{(q)_n} z^n = \frac{(bz)_\infty}{(z)_\infty}, \tag{4.50}$$

we take the sum over $\{n_j\}$ as

$$
\begin{aligned}
S &= \sum_{m_i} \prod_{i=1}^{M} \frac{(a_i)_{m_i}}{(q)_{m_i}} x_i^{m_i} \prod_{j=1}^{N} \frac{(b_j y_j q^{|m|})_\infty}{(y_j q^{|m|})_\infty} \\
&= \prod_{j=1}^{N} \frac{(b_j y_j)_\infty}{(y_j)_\infty} \sum_{m_i} \prod_{i=1}^{M} \frac{(a_i)_{m_i}}{(q)_{m_i}} x_i^{m_i} \prod_{j=1}^{N} \frac{(y_j)_{|m|}}{(b_j y_j)_{|m|}} \\
&= \prod_{j=1}^{N} \frac{(b_j y_j)_\infty}{(y_j)_\infty} \mathcal{F}_{N,M}\left(\begin{matrix} \{y_j\}, \{a_i\} \\ \{b_j y_j\} \end{matrix}; \{x_i\} \right).
\end{aligned} \tag{4.51}
$$

Similarly, summing over $\{m_i\}$ first, we have

$$S = \prod_{i=1}^{M} \frac{(a_i x_i)_\infty}{(x_i)_\infty} \mathcal{F}_{M,N}\left(\begin{matrix} \{x_i\}, \{b_j\} \\ \{a_i x_i\} \end{matrix}; \{y_j\} \right). \tag{4.52}$$

Hence (4.48) is proved.   □

Remark 4.4 In the case of $M = 1$ (or $N = 1$) the duality (4.48) reduces to the relation (4.39). In particular, the case $M = N = 1$ is *Heine's formula* for $_2\varphi_1$. The relation (4.48) can be obtained from a more general duality relation in [23].

Equation (4.53) below shows that the duality relation (4.48) can be considered as the *Jackson integral* representation for the function $\mathcal{F}_{M,N}$. In $q \to 1$ limit, this coincides with the integral representation of the function $F_{N+1,M}$ defined by Tsuda [61].

Proposition 4.6 *[49] Under the substitutions* $a_i = q^{\alpha_i}$, $b_j = q^{\beta_j}$ *and* $y_j = q^{\gamma_j}$, *(4.48) takes the following form:*

$$
\begin{aligned}
&\mathcal{F}_{N,M}\left(\begin{matrix} \{q^{\gamma_j}\}, \{q^{\alpha_i}\} \\ \{q^{\beta_j + \gamma_j}\} \end{matrix}; \{x_i\} \right) \\
&= \prod_{j=1}^{N} \frac{\Gamma_q(\beta_j + \gamma_j)}{\Gamma_q(\beta_j) \Gamma_q(\gamma_j)} \prod_{j=1}^{N} \int_0^1 d_q t_j \prod_{i=1}^{M} \frac{(x_i q^{\alpha_i} \prod_{j=1}^{N} t_j)_\infty}{(x_i \prod_{j=1}^{N} t_j)_\infty} \prod_{j=1}^{N} \frac{(q t_j)_\infty}{(q^{\beta_j} t_j)_\infty} t_j^{\gamma_j - 1},
\end{aligned} \tag{4.53}
$$

where q-Gamma function $\Gamma_q(z)$ and Jackson integral is defined as

$$\Gamma_q(z) = \frac{(q)_\infty}{(q^z)_\infty} (1-q)^{1-z}, \quad \int_0^c d_q t f(t) = c(1-q) \sum_{n=0}^{\infty} f(cq^n) q^n. \tag{4.54}$$

Proof Putting $a_i = q^{\alpha_i}$, $b_j = q^{\beta_j}$, $y_j = q^{\gamma_j}$, the RHS of (4.48) is written as

$$
\frac{(\{q^{\gamma_j}\}, \{q^{\alpha_i} x_i\})_\infty}{(\{q^{\beta_j+\gamma_j}\}, \{x_i\})_\infty} \sum_{n_j} \prod_{i=1}^{M} \frac{(x_i)_{|n|}}{(q^{\alpha_i} x_i)_{|n|}} \prod_{j=1}^{N} \frac{(q^{\beta_j})_{n_j}}{(q)_{n_j}} q^{\gamma_j n_j}
$$

$$
= \frac{(\{q^{\gamma_j}\})_\infty}{(\{q^{\beta_j+\gamma_j}\})_\infty} \prod_{j=1}^{N} \frac{(q^{\beta_j})_\infty}{(q)_\infty} \sum_{n_j} \prod_{i=1}^{M} \frac{(x_i q^{\alpha_i+|n|})_\infty}{(x_i q^{|n|})_\infty} \prod_{j=1}^{N} \frac{(q^{1+n_j})_\infty}{(q^{\beta_j+n_j})_\infty} (q^{n_j})^{\gamma_j}
$$

$$
= \frac{(\{q^{\gamma_j}\})_\infty}{(\{q^{\beta_j+\gamma_j}\})_\infty} \prod_{j=1}^{N} \frac{(q^{\beta_j})_\infty}{(q)_\infty} \prod_{j=1}^{N} \int_0^1 \frac{d_q t_j}{1-q} \prod_{i=1}^{M} \frac{(x_i q^{\alpha_i} t_1 \cdots t_N)_\infty}{(x_i t_1 \cdots t_N)_\infty} \prod_{j=1}^{N} \frac{(q t_j)_\infty}{(q^{\beta_j} t_j)_\infty} t_j^{\gamma_j-1}
$$

$$
= \prod_{j=1}^{N} \frac{\Gamma_q(\beta_j + \gamma_j)}{\Gamma_q(\beta_j)\Gamma_q(\gamma_j)} \prod_{j=1}^{N} \int_0^1 d_q t_j \prod_{i=1}^{M} \frac{(x_i q^{\alpha_i} t_1 \cdots t_N)_\infty}{(x_i t_1 \cdots t_N)_\infty} \prod_{j=1}^{N} \frac{(q t_j)_\infty}{(q^{\beta_j} t_j)_\infty} t_j^{\gamma_j-1}.
$$

$$(4.55)$$

Hence we obtain (4.53). $\qquad\square$

Chapter 5
Padé Interpolation on q-Quadratic Grid

Abstract In the Padé interpolation, one can make various choices for the grid (interpolating points: $x_0, x_1, x_2, \cdots$). Here we discuss applications of the Padé interpolation on a q-quadratic grid defined by $x_s = q^s + cq^{-s}$.

5.1 Contiguous Relations

For $0 < |q| < 1$, we use the following notations as before:

$$(x)_\infty := \prod_{i=0}^{\infty} (1 - q^i x), \quad (x)_s := \frac{(x)_\infty}{(xq^s)_\infty}, \quad (x_1, x_2, \ldots, x_k)_s := (x_1)_s (x_2)_s \ldots (x_k)_s.$$

$$(5.1)$$

For a given sequence ψ_s ($s = 0, 1, 2, \cdots$), consider a Padé interpolation problem

$$\psi_s = \frac{P(x_s)}{Q(x_s)} \quad (s = 0, 1, \ldots m + n), \tag{5.2}$$

where $P(x)$ and $Q(x)$ are polynomials in x of degree m and n, and the interpolation points $\{x_s\}$ are taken as a *q-quadratic grid*, i.e.

$$x_s = z_s + \frac{qk}{z_s}, \quad z_s = q^s. \tag{5.3}$$

In the following, we choose the sequence $\psi_s = \psi(q^s)$ as

© The Author(s), under exclusive license to Springer Nature Singapore Pte Ltd 2021
H. Nagao and Y. Yamada, *Padé Methods for Painlevé Equations*,
SpringerBriefs in Mathematical Physics,
https://doi.org/10.1007/978-981-16-2998-3_5

$$\psi_s = \prod_{i=1}^{N} \left(\frac{a_i}{b_i}\right)^s \frac{(\frac{1}{ka_i}, b_i)_s}{(a_i, \frac{1}{kb_i})_s}, \quad \psi(z) = \prod_{i=1}^{N} \frac{(a_i z, \frac{qka_i}{z}, b_i, qkb_i)_\infty}{(a_i, qka_i, b_i z, \frac{qkb_i}{z})_\infty}, \quad (5.4)$$

where $k, a_1, \ldots, a_N, b_1, \ldots, b_N$ are complex parameters.

It is convenient to use a variable z such that $x = z + \frac{qk}{z}$, and introduce Laurent polynomials in z by $\mathcal{P}(z) = P(x), \mathcal{Q}(z) = Q(x)$, i.e.

$$\mathcal{P}(z) = \sum_{i=0}^{m} u_i (z + \frac{qk}{z})^i, \quad \mathcal{Q}(z) = \sum_{i=0}^{n} v_i (z + \frac{qk}{z})^i. \quad (5.5)$$

Note that the function $\psi(z)$ and the polynomials $\mathcal{P}(z)$ and $\mathcal{Q}(z)$ are qk-symmetric:

$$\psi(\frac{qk}{z}) = \psi(z), \quad \mathcal{P}(\frac{qk}{z}) = \mathcal{P}(z), \quad \mathcal{Q}(\frac{qk}{z}) = \mathcal{Q}(z). \quad (5.6)$$

In other words, the interpolation on a q-quadratic grid can be considered as an interpolation on a q-grid with a symmetry constraint.

For the time evolution, we consider a q-shift T corresponding to the direction

$$T : \{k, a_i, b_i, m, n\} \to \{q^{-1}k, a_i, b_i, m, n\}, \quad (5.7)$$

and use the notation $\overline{x} = T(x)$ and $\underline{x} = T^{-1}(x)$.

We define the following functions:

$$M(z) = \frac{M_{\mathrm{num}}(z)}{M_{\mathrm{den}}(z)} = \frac{\psi(qz)}{\psi(z)}, \quad K(z) = \frac{K_{\mathrm{num}}(z)}{K_{\mathrm{den}}(z)} = \frac{\overline{\psi}(z)}{\psi(z)}, \quad (5.8)$$

$$M_{\mathrm{den}}(z) = \prod_{i=1}^{N} (a_i z, \frac{b_i k}{z})_1, \quad M_{\mathrm{num}}(z) = \prod_{i=1}^{N} (\frac{a_i k}{z}, b_i z)_1,$$

$$K_{\mathrm{den}}(z) = \prod_{i=1}^{N} (a_i k, \frac{b_i k}{z})_1, \quad K_{\mathrm{num}}(z) = \prod_{i=1}^{N} (\frac{a_i k}{z}, b_i k)_1,$$

$$H(z) = \prod_{i=1}^{N} (a_i z, a_i k, \frac{b_i k}{z})_1,$$

where $H(z)$ is the least common multiple of $M_{\mathrm{den}}(z), K_{\mathrm{den}}(z)$. Then we have relations

$$M(\frac{k}{z})M(z) = 1, \quad K(z) = M(z)K(\frac{k}{z}). \quad (5.9)$$

Let us consider two linear difference relations of contiguous type: L_2 between $y(z), y(qz), \overline{y}(z)$ and L_3 between $y(z), \overline{y}(z), \overline{y}(\frac{z}{q})$, whose solutions are the functions $y(z) = \mathcal{P}(z)$ and $y(z) = \psi(z)Q(z)$. Then, the linear relations are expressed as:

$$L_2: \quad D_1(z)\overline{y}(z) - D_2(z)y(qz) + D_3(z)y(z) = 0, \tag{5.10}$$

$$L_3: \quad \overline{D}_1(\frac{z}{q})y(z) + D_3(\frac{z}{q})\overline{y}(z) - D_2(z)\overline{y}(\frac{z}{q}) = 0.$$

Here the coefficients of y and $\overline{y}$ are determined by Casorati determinants

$$D_1(z) = |\mathbf{u}(z), \mathbf{u}(qz)|, \quad D_2(z) = |\mathbf{u}(z), \overline{\mathbf{u}}(z)|, \tag{5.11}$$

$$D_3(z) = |\mathbf{u}(qz), \overline{\mathbf{u}}(z)|, \quad \mathbf{u}(z) = \begin{bmatrix} \mathcal{P}(z) \\ \psi(z)Q(z) \end{bmatrix}.$$

Lemma 5.1 *The Casorati determinant $D_1(z)$ in (5.11) is written as*

$$D_1(z) = \frac{\psi(z)}{M_{\mathrm{den}}(z)} \prod_{i=0}^{m+n-1} (\frac{z}{q^i}, \frac{k}{q^i z})_1 w(z) F(z), \tag{5.12}$$

where

$$F(z) = \sum_{i=0}^{N-1} f_i \left(z + \frac{k}{z}\right)^i, \quad w(z) = z - \frac{k}{z}, \tag{5.13}$$

and $f_0, \ldots, f_{N-1}$ are some constants.

Proof We put

$$D_1(z) = \psi(qz)\mathcal{P}(z)Q(qz) - \psi(z)\mathcal{P}(qz)Q(z) = \frac{\psi(z)}{M_{\mathrm{den}}(z)} E_1(z), \tag{5.14}$$

$$E_1(z) = M_{\mathrm{num}}(z)\mathcal{P}(z)Q(qz) - M_{\mathrm{den}}(z)\mathcal{P}(qz)Q(z).$$

It is easy to see the following:

(i) $E_1(z)$ is a *Laurent polynomial* of degree $(-m - n - N, m + n + N)$ in z, i.e.
$E_1(z) = \sum_{i=-m-n-N}^{m+n+N} e_i z^i$.

(ii) Thanks to the relations $M_{\mathrm{num}}(\frac{k}{z}) = M_{\mathrm{den}}(z)$, $M_{\mathrm{den}}(\frac{k}{z}) = M_{\mathrm{num}}(z)$, we have

$$E_1(\frac{k}{z}) = M_{\mathrm{num}}(\frac{k}{z})\mathcal{P}(\frac{k}{z})Q(\frac{qk}{z}) - M_{\mathrm{den}}(\frac{k}{z})\mathcal{P}(\frac{qk}{z})Q(\frac{k}{z}) \tag{5.15}$$

$$= M_{\mathrm{den}}(z)\mathcal{P}(qz)Q(z) - M_{\mathrm{num}}(z)\mathcal{P}(z)Q(qz) = -E_1(z).$$

By (ii) and the Padé interpolation condition (5.2), $E_1(z)$ is factorized by $\prod_{i=0}^{m+n-1}(\frac{z}{q^i}, \frac{k}{q^i z})_1 w(z)$. The remaining factor $F(z)$ is a Laurent polynomial of degree $(-N + 1, N - 1)$ in z such that $F(\frac{k}{z}) = F(z)$. Hence, we obtain the desired result. $\square$

Lemma 5.2 *The Casorati determinants $D_2(z)$ and $D_3(z)$ in (5.11) are written as*

$$D_2(z) = \frac{\psi(z)}{K_{\mathrm{den}}(z)} \prod_{i=0}^{m+n} (\frac{z}{q^i})_1 \prod_{i=0}^{m+n-1} (\frac{k}{q^i z})_1 G(\frac{k}{z}), \tag{5.16}$$

$$D_3(z) = \frac{\psi(z)}{H(z)} \prod_{i=0}^{m+n-1} (\frac{z}{q^i})_1 \prod_{i=0}^{m+n} (\frac{k}{q^i z})_1 \prod_{i=1}^{N} (\frac{a_i k}{z})_1 G(z), \tag{5.17}$$

where

$$G(z) = z \sum_{i=0}^{N-1} g_i z^i, \tag{5.18}$$

and $g_0, \ldots, g_{N-1}$ are some constants.

Proof From (5.6) and its T-shift

$$\overline{\psi}(\frac{k}{z}) = \overline{\psi}(z), \quad \overline{\mathcal{P}}(\frac{k}{z}) = \overline{\mathcal{P}}(z), \quad \overline{Q}(\frac{k}{z}) = \overline{Q}(z), \tag{5.19}$$

we have the following relation between D_2 and D_3:

$$\begin{aligned}
D_3(\frac{k}{z}) &= \overline{\psi}(\frac{k}{z})\mathcal{P}(\frac{qk}{z})\overline{Q}(\frac{k}{z}) - \psi(\frac{qk}{z})\overline{\mathcal{P}}(\frac{k}{z})Q(\frac{qk}{z}) \tag{5.20} \\
&= \overline{\psi}(\frac{k}{z})\mathcal{P}(z)\overline{Q}(z) - \psi(\frac{qk}{z})\overline{\mathcal{P}}(z)Q(z) \\
&= \overline{\psi}(z)\mathcal{P}(z)\overline{Q}(z) - \psi(z)\overline{\mathcal{P}}(z)Q(z) = D_2(z).
\end{aligned}$$

From this and the interpolation condition, we see that $D_2(z)$ (resp. $D_3(z)$) is divisible by $\prod_{i=0}^{m+n} (\frac{z}{q^i})_1 \prod_{i=0}^{m+n-1} (\frac{k}{q^i z})_1$ (resp. $\prod_{i=0}^{m+n-1} (\frac{z}{q^i})_1 \prod_{i=0}^{m+n} (\frac{k}{q^i z})_1$). Hence, we obtain the desired result (5.16) (resp. (5.17)). $\qquad\square$

Proposition 5.1 *The linear relations L_2 and L_3 can be expressed as follows:*

$$L_2: \quad w(z)F(z)\overline{y}(z) - A(z)G(\frac{k}{z})y(qz) + A(\frac{k}{z})G(z)y(z) = 0, \tag{5.21}$$

$$L_3: \quad \overline{w}(\frac{z}{q})\overline{F}(\frac{z}{q})y(z) + B(\frac{k}{z})G(\frac{z}{q})\overline{y}(z) - B(\frac{z}{q})G(\frac{k}{z})\overline{y}(\frac{z}{q}) = 0,$$

where

$$F(z) = \sum_{i=0}^{N-1} f_i \left(z + \frac{k}{z}\right)^i, \quad G(z) = z \sum_{i=0}^{N-1} g_i z^i, \quad w(z) = z - \frac{k}{z}, \tag{5.22}$$

$$A(z) = (\frac{z}{q^{m+n}})_1 \prod_{i=1}^{N} (a_i z)_1, \quad B(z) = (qz)_1 \prod_{i=1}^{N} (b_i z)_1.$$

Proof Substituting the results in Lemmas 5.1 and 5.2 into the linear relations (5.10), one obtain the desired equations.

Proposition 5.2 *The pair of equations L_2 and L_3 (5.21) gives the following relations as necessary conditions for the compatibility:*

$$\frac{G(z)\underline{G}(z)}{G(\frac{k}{z})\underline{G}(\frac{k}{z})} = \frac{A(z)B(z)}{A(\frac{k}{z})B(\frac{k}{z})} \quad \text{for} \quad F(z) = 0, \tag{5.23}$$

$$w(z)\overline{w}(z)F(z)\overline{F}(z) = A(z)B(z)G(\frac{k}{z})G(\frac{k}{qz}) \quad \text{for} \quad G(z) = 0, \tag{5.24}$$

$$f_{N-1}\overline{f}_{N-1} = \left\{ g_{N-1} - (\frac{k}{q^n} \prod_{s=1}^{N} a_s)g_0 \right\}\left\{ g_{N-1} - (\frac{k}{q^m} \prod_{s=1}^{N} b_s)g_0 \right\}. \tag{5.25}$$

Proof Under the condition $F(z) = 0$, L_2 and $\underline{L}_3|_{z \to qz}$ give

$$-A(z)G(\frac{k}{z})y(qz) + A(\frac{k}{z})G(z)y(z) = 0, \tag{5.26}$$

$$B(\frac{k}{z})G(z)y(qz) - B(z)\underline{G}(\frac{k}{z})y(z) = 0.$$

Then we obtain equation (5.23) by eliminating $y(z)$ and $y(qz)$. Similarly, under the condition $G(z) = 0$, we have (5.24) from L_2 and $L_3|_{z \to qz}$. Considering the solutions $y(z) = P(z)$ around $z = 0$ and $z = \infty$ we obtain (5.25). $\qquad\square$

5.2 Lax Pair and the Compatibility

For $y(z) = P(z), \psi(z)Q(z)$, we have the linear equation L_1 between $y(qz)$, $y(z)$, $y(z/q)$:

$$L_1 : A(z)B(\frac{k}{z})w(\frac{z}{q})F(\frac{z}{q})y(qz) + A(\frac{qk}{z})B(\frac{z}{q})w(z)F(z)y(\frac{z}{q}) - R(z)y(z) = 0. \tag{5.27}$$

An explicit form of the coefficient $R(z)$ can be derived from (5.21). Eliminating $\overline{y}(z)$ and $\overline{y}(\frac{z}{q})$ from L_2, $L_2|_{z \to \frac{z}{q}}$, L_3, we have

$$R(z) = \frac{A(\frac{z}{q})B(\frac{z}{q})w(z)F(z)G(\frac{qk}{z})}{G(\frac{z}{q})} + \frac{A(\frac{k}{z})B(\frac{k}{z})w(\frac{z}{q})F(\frac{z}{q})G(z)}{G(\frac{k}{z})}$$

$$- \frac{w(z)F(z)w(\frac{z}{q})F(\frac{z}{q})\overline{w}(\frac{z}{q})\overline{F}(\frac{z}{q})}{G(\frac{z}{q})G(\frac{k}{z})}. \tag{5.28}$$

Lemma 5.3 *The expression $R(z)$ (5.28) has the following properties:*
(i) $R(z)$ is a Laurent polynomial of degree $(-2N - 1, 2N + 1)$ in z.
(ii) $R(z)$ has a symmetry $R(\frac{qk}{z}) = -R(z)$.
(iii) The exponents of solutions $y(z)$ of the equation L_1 are $\frac{1}{c_1}, \frac{1}{c_2}$ (at $z = 0$) and c_1,
c_2 *(at $z = \infty$), where $c_1 = q^m$, $qc_1c_2 = \prod_{i=1}^{N+1} \frac{b_i}{a_i}$, $a_{N+1} = q^{-(m+n)}$ and $b_{N+1} = q$.*
(iv) The equation L_1 holds when

$$\frac{y(qz)}{y(z)} = \frac{A(\frac{k}{z})G(z)}{A(z)G(\frac{k}{z})}, \quad w(z)F(z) = 0, \tag{5.29}$$

$$\text{or} \quad \frac{y(z)}{y(\frac{z}{q})} = \frac{A(\frac{qk}{z})G(\frac{z}{q})}{A(\frac{z}{q})G(\frac{qk}{z})}, \quad w(\frac{z}{q})F(\frac{z}{q}) = 0.$$

Conversely, if the coefficients of $y(qz)$ and $y(\frac{z}{q})$ for the three terms relation L_1 are fixed as in (5.27), then the coefficient $R(z)$ is uniquely determined from the properties (i)–(iv).

Proof $R(z)$ (5.28) can be rewritten as $\frac{R_{num}(z)}{R_{den}(z)}$, where

$$R_{num}(z) = A(\frac{z}{q})B(\frac{z}{q})w(z)F(z)G(\frac{qk}{z})G(\frac{k}{z}) + A(\frac{k}{z})B(\frac{k}{z})w(\frac{z}{q})F(\frac{z}{q})G(z)G(\frac{z}{q})$$
$$- w(z)F(z)w(\frac{z}{q})F(\frac{z}{q})\overline{w}(\frac{z}{q})\overline{F}(\frac{z}{q}), \tag{5.30}$$

$$R_{den}(z) = G(\frac{z}{q})G(\frac{k}{z}). \tag{5.31}$$

The numerator $R_{num}(z)$ (resp. the denominator $R_{den}(z)$) is a Laurent polynomial of degree $(-3N - 2, 3N + 2)$ (resp. $(-N - 1, N + 1)$) in z, and satisfies $R_{num}(\frac{qk}{z}) = -R_{num}(z)$ (resp. $R_{den}(\frac{qk}{z}) = R_{den}(z)$). Hence, the property (ii) is confirmed. Thanks to the equation (5.24), we see that $R_{num}(z)$ is divisible by the Laurent polynomial $G(\frac{z}{q})G(\frac{k}{z})$. The polynomial $G(\frac{z}{q})G(\frac{k}{z})$ is of degree $(-N - 1, N + 1)$ in z. Therefore, the property (i) is proved. The property (iii) follows from the consideration of the limit $z \to 0$ ($z \to \infty$). The property (iv) is obvious. The uniqueness follows from the counting the parameters and conditions. $\square$

Similarly, eliminating $y(z)$ and $y(qz)$ from $L_2, L_3, L_3|_{z \to qz}$, we have the following linear equation:

$$L_1^* : A(z)B(\frac{k}{qz})\overline{w}(\frac{z}{q})\overline{F}(\frac{z}{q})\overline{y}(qz) + A(\frac{k}{z})B(\frac{z}{q})\overline{w}(z)\overline{F}(z)y(\frac{z}{q}) - R^*(z)\overline{y}(z) = 0, \tag{5.32}$$

where

$$R^*(z) = \frac{A(\frac{k}{z})B(\frac{z}{q})\overline{w}(z)\overline{F}(z)G(\frac{z}{q})}{G(\frac{k}{z})} + \frac{A(z)B(z)\overline{w}(\frac{z}{q})\overline{F}(\frac{z}{q})G(\frac{k}{qz})}{G(z)}$$

$$- \frac{w(z)F(z)\overline{w}(z)\overline{F}(z)\overline{w}(\frac{z}{q})\overline{F}(\frac{z}{q})}{G(z)G(\frac{k}{z})}. \tag{5.33}$$

The following can be proved in the same way as Lemma 5.3.

Lemma 5.4 *The equation $R^*(z) = 0$ (5.33) has the following properties:*
(i) $R^*(z)$ is a Laurent polynomial of degree $(-2N - 1, 2N + 1)$ in z.
(ii) $R^*(\frac{k}{z}) = -R^*(z)$.
(iii) The exponents of solutions $\overline{y}(z)$ of the equation L_1^* are $\frac{1}{c_1}, \frac{1}{c_2}$ (at $z = 0$) and c_1, c_2 (at $z = \infty$).
(iv) The equation L_1^* holds when

$$\frac{\overline{y}(qz)}{\overline{y}(z)} = \frac{B(z)G(\frac{k}{qz})}{B(\frac{k}{qz})G(z)}, \quad \overline{w}(z)\overline{F}(z) = 0, \tag{5.34}$$

$$\text{or} \quad \frac{\overline{y}(z)}{\overline{y}(\frac{z}{q})} = \frac{B(\frac{z}{q})G(\frac{k}{z})}{B(\frac{k}{z})G(\frac{z}{q})}, \quad \overline{w}(\frac{z}{q})\overline{F}(\frac{z}{q}) = 0.$$

Furthermore, the coefficient $R^*(z)$ is uniquely characterized by properties (i)–(iv).

For the off-Padé situation, the equations L_2, L_3 (or equivalently L_2, L_1) can be viewed as the Lax pair for certain isomonodromic deformations. In fact, one can prove that the equations (5.23)–(5.25) are sufficient conditions for the compatibility of the Lax pair L_2, L_3, using the characterization of L_1 and L_1^* given above (see [41]).

Example

For the case $N = 2$, putting $f = -\frac{f_0}{f_1}$, $g = -\frac{g_0}{g_1}$, the equations (5.23)–(5.25) takes the following form:

$$\frac{(g - z)(g - z)}{(g - \frac{k}{z})(g - \frac{k}{z})} = \frac{(\frac{k}{z})^2 A(z)B(z)}{z^2 A(\frac{k}{z})B(\frac{k}{z})} \quad \text{for} \quad f - \left(z + \frac{k}{z}\right) = 0, \tag{5.35}$$

$$\left\{f - \left(g + \frac{k}{g}\right)\right\}\left\{\overline{f} - \left(g + \frac{k}{qg}\right)\right\} = \frac{A(g)B(g)}{g^2(g + \frac{qc_1B(0)}{k})(g + \frac{qc_2B(0)}{k})},$$

where $qc_1c_2 = \prod_{i=1}^3 \frac{a_i}{b_i}$. The equations (5.35) are equivalent to the equations (4.25) (a non-standard version of q-*Painlevé equation of type* $E_7^{(1)}$ [38, 40, 41]).

For the case $N = 3$ with an constraint $c_1 = c_2$, (i.e. $q^{m-n} \prod_{i=1}^3 \frac{a_i}{b_i} = 1$), the equations (5.23)–(5.25) admit a specialization $f_2 = 0$ and $g_0 = lg_2$, where $kl \prod_{i=1}^4 a_i =$

q^n, $a_4 = q^{-(m+n)}$ and $b_4 = q$. Then by a variable transformation $f = -\frac{f_0}{f_1}$, $g = -\frac{g_1}{g_2}$, we have the following equations [65]:

$$\frac{G(z)\underline{G}(z)}{G(\frac{k}{z})\underline{G}(\frac{k}{z})} = \frac{U(z)}{U(\frac{k}{z})} \quad \text{for} \quad F(z) = 0, \tag{5.36}$$

$$\frac{F(z)\overline{F}(z)}{F(\frac{l}{z})\overline{F}(\frac{l}{z})} = \frac{U(z)}{U(\frac{l}{z})} \quad \text{for} \quad G(z) = 0,$$

with $\overline{k} = kq^{-1}$, $\overline{l} = q\ell$, and

$$F(z) = f - (z + \frac{k}{z}), \quad G(z) = g - (z + \frac{l}{z}), \quad U(z) = \frac{1}{z^4}\prod_{i=1}^{4}(z - a_i)(z - b_i).$$
$$\tag{5.37}$$

Note that $F(z)$, $G(z)$ are rescaled as $F(z) = -f_1 F(z)$, $G(z) = -g_2 z^2 G(z)$ to simplify the notation. The system of equations (5.36) is known as the q-*Painlevé equation of type* $E_8^{(1)}$ (see [25] and references therein). From (5.36) the variables $\overline{f}$, g can be expressed as rational functions of f, g. See [44] for the explicit rational expression.

5.3 Special Solutions

First we recall Jacobi's formula (4.6) for the interpolation problem

$$P(x) = f(x) \det\left[\mathcal{W}_{i,j}^{(-)}(x)\right]_{i,j=0}^{n}, \quad Q(x) = \det\left[\mathcal{W}_{i,j}^{(+)}(x)\right]_{i,j=0}^{n-1}, \tag{5.38}$$

where

$$\mathcal{W}_{i,j}^{(\pm)}(x) = \sum_{s=0}^{m+n} \frac{\psi_s}{f'(x_s)} x_s^{i+j}(x - x_s)^{(\pm 1)}. \tag{5.39}$$

Lemma 5.5 *For the q-quadratic grid case, i.e., $x = x(z) = z + \frac{qk}{z}$, $x_s = x(q^s)$, we have*

$$f(x) = (-1)^{m+n+1}q^{-\binom{m+n+1}{2}}(qk)^{m+n+1}(\frac{1}{z}, \frac{z}{qk})_{m+n+1}, \tag{5.40}$$

$$f'(x_s) = \frac{(-1)^{m+n}(qk)^{m+n}(q, \frac{1}{k})_{m+n}}{q^{\binom{m+n+1}{2}}} \frac{(\frac{1}{qk})_1(q, \frac{q^{m+n}}{k})_s}{(\frac{q^{2s-1}}{k})_1(\frac{1}{qk}, q^{-(m+n)})_s} q^{-(m+n)s}, \tag{5.41}$$

where $\binom{a}{2} = \frac{a(a-1)}{2}$.

Proof From a relation $x(u) - x(v) = -\dfrac{qk}{v}(\dfrac{v}{u}, \dfrac{uv}{qk})_1$, we have

$$f(x) = \prod_{s=0}^{m+n}(x - x_s) = \prod_{s=0}^{m+n}(-\frac{qk}{q^s})(\frac{q^s}{z}, \frac{q^s z}{qk})_1,\qquad (5.42)$$

and hence (5.40). Similarly we have (5.41) from

$$f'(x_s) = \lim_{x \to x_s} \frac{f(x)}{x - x_s} = \lim_{z \to q^s} \prod_{t(\ne s)=0}^{m+n}(-\frac{qk}{q^t})(\frac{q^t}{z}, \frac{q^t z}{qk})_1.\qquad (5.43)$$

$\square$

Lemma 5.6 *In the determinants (5.38), the elements* $\mathcal{W}_{i,j}^{(\pm)}(x)$ *can be replaced as*

$$\mathcal{W}_{i,j}^{(\pm)}(x) = (kq)^{i+j}\mathbf{g}^{-1}\sum_{s=0}^{m+n}\frac{T_{a_N}^{-i}T_{b_N}^{j}(\mathbf{g}\psi_s)}{f'(x_s)}(x - x_s)^{(\pm1)},\qquad (5.44)$$

where

$$\mathbf{g} = \prod_{i=1}^{N}\frac{(a_i, \frac{1}{kb_i})_\infty}{(b_i, \frac{1}{ka_i})_\infty}.\qquad (5.45)$$

Proof Recall that the factors x_s^i (or x_s^j) in Jacobi's determinant formula can be replaced by any monic polynomial in x_s of degree i (or j) (see Remark 4.1). An example of convenient choices[1] is

$$\psi_s x_s^i x_s^j \to \psi_s \prod_{\mu=1}^{i}\left\{x_s - x(\frac{q^\mu}{a_N})\right\}\prod_{v=0}^{j-1}\left\{x_s - x(\frac{q^v}{b_N})\right\}.\qquad (5.46)$$

Then, noting the relations

$$T_{a_i}(\mathbf{g}\psi(z)) = \mathbf{g}\psi(z)\frac{kq}{x(z) - x(\frac{1}{a_i})},\quad T_{a_i}^{-1}(\mathbf{g}\psi(z)) = \mathbf{g}\psi(z)\frac{x(z) - x(\frac{q}{a_i})}{kq}, (5.47)$$

$$T_{b_i}(\mathbf{g}\psi(z)) = \mathbf{g}\psi(z)\frac{x(z) - x(\frac{1}{b_i})}{kq},\quad T_{b_i}^{-1}(\mathbf{g}\psi(z)) = \mathbf{g}\psi(z)\frac{kq}{x(z) - x(\frac{q}{b_i})}, (5.48)$$

we obtain the result.

$\square$

Define the τ functions as

---

[1] Here we have chosen the the parameter a_N (or b_N) as an example. However, other choices are also possible. Accordingly, in (5.44) and (5.49), the shift T_{a_N} (or T_{b_N}) in the RHS can be replaced by any other T_{a_k} (or T_{b_l}) without changing the LHS.

$$\tau_{m,n} = \det\left[T_{a_N}^{-i} T_{b_N}^{j}\left(U \,_{2N+4}W_{2N+3}(u_1, \ldots, u_{2N+2}; x)\right)\right]_{i,j=0}^{n}, \qquad (5.49)$$

where $_{\nu+2}W_{\nu+1}$ is the *very-well-poised q-hypergeometric series* [10, 57]

$$_{\nu+2}W_{\nu+1}(u_1, u_2, \ldots, u_\nu; x) = \sum_{s=0}^{\infty} \frac{(u_1 q^{2s})_1}{(u_1)_1}\frac{(u_1, \ldots, u_\nu)_s}{(q, \frac{qu_1}{u_2}, \ldots, \frac{qu_1}{u_\nu})_s} x^s, \quad (5.50)$$

$$\{u_1, \ldots, u_{2N+2}\} = \{\frac{1}{qk}, \frac{1}{q^{m+n}}, \frac{1}{ka_1}, \ldots, \frac{1}{ka_N}, b_1, \ldots, b_N\}, \qquad (5.51)$$

$$x = q^{m+n}\prod_{r=1}^{N}\frac{a_r}{b_r}, \quad U = \frac{\mathbf{g}}{(q, \frac{1}{k})_{m+n}}. \qquad (5.52)$$

Proposition 5.3 *The polynomial $P(x) = \mathcal{P}(z)$ (resp. $Q(x) = \mathcal{Q}(z)$) for the interpolation problem in Sect. 5.1 has the following special values at $z = \frac{1}{a_i}, \frac{q}{b_i}$ (resp. $z = \frac{q}{a_i}, \frac{1}{b_i}$) for $i = 1, \ldots, N$:*

$$\mathcal{P}(\frac{1}{a_i}) = (a_i, \frac{1}{qka_i})_{m+n+1}\rho\mathbf{g}^{-n-1}T_{a_i}(\tau_{m,n}), \qquad (5.53)$$

$$\mathcal{P}(\frac{q}{b_i}) = (\frac{b_i}{q}, \frac{1}{kb_i})_{m+n+1}\rho\mathbf{g}^{-n-1}T_{b_i}^{-1}(\tau_{m,n}), \qquad (5.54)$$

$$\mathcal{Q}(\frac{q}{a_i}) = \rho\mathbf{g}^{-n}T_{a_i}^{-1}(\tau_{m+1,n-1}), \qquad (5.55)$$

$$\mathcal{Q}(\frac{1}{b_i}) = \rho\mathbf{g}^{-n}T_{b_i}(\tau_{m+1,n-1}), \qquad (5.56)$$

$$\rho = (-1)^{(m+n+1)n}q^{\binom{m+n+1}{2}n}(qk)^{-mn}. \qquad (5.57)$$

Proof The extra factor $(x - x_s)^{\pm 1}$ in $\mathcal{W}_{i,j}^{(\pm)}(x)$ can also be absorbed as a shift of parameter if we specialize $x = x(z)$ suitably. Then, applying Lemmas 5.5 and 5.6, we obtain the desired results. $\qquad\square$

Proposition 5.4 *Define the polynomials $F(z)$ and $G(z)$ by the following values at the special points at $z = \frac{1}{a_i}, \frac{1}{b_j}$ $(i, j=1,\ldots, N)$:*

$$\frac{F(\frac{1}{a_i})}{F(\frac{1}{b_j})} = \alpha \frac{T_{a_i}(\tau_{m,n})T_{a_i}^{-1}(\tau_{m+1,n-1})}{T_{b_j}^{-1}(\tau_{m,n})T_{b_j}(\tau_{m+1,n-1})}, \tag{5.58}$$

$$\frac{G(\frac{1}{a_i})}{G(\frac{1}{b_j})} = \beta \frac{T_{a_i}T_k^{-1}(\tau_{m,n})T_{a_i}^{-1}(\tau_{m+1,n-1})}{T_{b_j}^{-1}(\tau_{m,n})T_{b_j}T_k^{-1}(\tau_{m+1,n-1})}, \tag{5.59}$$

$$\alpha = -\frac{a_i}{b_j} \frac{(a_1 q^{m+n}, \frac{1}{qka_i}, kb_j^2)_1 \prod_{s=1}^{N}(ka_i a_s, \frac{b_s}{a_i})_1}{(\frac{q^{m+n}}{kb_j}, \frac{b_j}{q}, ka_i^2)_1 \prod_{s=1}^{N}(kb_j b_s, \frac{a_s}{b_j})_1}, \tag{5.60}$$

$$\beta = -\frac{b_j}{a_i} \frac{(a_i q^{m+n})_1 \prod_{s=1}^{N} a_s(\frac{b_s}{a_i})_1}{(\frac{b_j}{q})_1 \prod_{s=1}^{N} b_s(\frac{a_s}{b_j})_1}, \tag{5.61}$$

then they satisfy the evolution equation in Proposition 5.2.

Proof Comparing the ratio $\frac{D_1(\frac{1}{a_i})}{D_1(\frac{1}{b_j})}$ obtained from (5.12) and (5.14), we have

$$\frac{F(\frac{1}{a_i})}{F(\frac{1}{b_j})} = -\frac{\frac{1}{b_j} - kb_j}{\frac{1}{a_i} - ka_i} \frac{M_{\text{num}}(\frac{1}{a_i})}{M_{\text{den}}(\frac{1}{b_j})} \frac{(b_j, \frac{1}{kb_j})_{m+n}}{(a_i, \frac{1}{ka_i})_{m+n}} \frac{P(\frac{1}{a_i})Q(\frac{q}{a_i})}{P(\frac{q}{b_j})Q(\frac{1}{b_j})}. \tag{5.62}$$

Comparing the ratio $\frac{D_3(\frac{1}{a_i})}{D_3(\frac{1}{b_j})}$ from (5.16) and (5.20), we get

$$\frac{G(\frac{1}{a_i})}{G(\frac{1}{b_j})} = -\frac{b_j(b_j)_{m+n}(\frac{1}{kb_j})_{m+n+1}}{a_i(a_i)_{m+n}(\frac{1}{ka_i})_{m+n+1}} \prod_{t=0}^{N} \frac{(b_j ka_t, ka_t)_1}{(a_i ka_t, \frac{a_t}{b_j})_1} \frac{M_{\text{num}}(\frac{1}{a_i})}{K_{\text{num}}(\frac{1}{b_j})} \frac{\overline{P}(\frac{1}{a_i})Q(\frac{q}{a_i})}{P(\frac{q}{b_j})\overline{Q}(\frac{1}{b_j})}. \tag{5.63}$$

Substituting the special values (5.53) into the expressions (5.62) and (5.63) respectively, we obtain the desired results. $\qquad\square$

For the q-Painlevé equation of type $E_8^{(1)}$ (see Example in Section 5.2), a special solution written as determinants of $_{10}W_9$ series has been obtained in [33] by using the bilinear formalism.

Other interpolation grids

The most generic interpolation grid $\{x_s\}$ which has been considered in relation with isomonodromic systems is the *elliptic grid* [68]. In the multiplicative notation, the elliptic grid can be written as

$$x_s = \varphi(z_s), \quad z_s = q^s, \quad \varphi(z) = \frac{\vartheta(az)\vartheta(bz)}{\vartheta(cz)\vartheta(dz)}, \tag{5.64}$$

where $\vartheta(z)$ is the theta function with base p,

$$\vartheta(z) = (z, p/z; p)_\infty = \prod_{n=0}^{\infty} (1 - zp^n)(1 - z^{-1}p^{n+1}). \qquad (5.65)$$

Thanks to the relation $\vartheta(pz) = -z^{-1}\vartheta(z)$, the function $\varphi(z)$ is p-periodic under the condition $ab/cd = 1$. In Sakai's geometric classification of discrete Painlevé equations, the *elliptic-$E_8^{(1)}$ Painlevé equation* is the most generic case [44, 51]. The Padé interpolation on elliptic grid has applications to the elliptic-$E_8^{(1)}$ Painlevé equation and its multivariate generalization [42, 66] (see also [47]).

Taking a limit $p \to 0$ and further specializing $d = 0, c \to \infty$, the function $\varphi(z)$ degenerates to

$$\varphi(z) \to \frac{(1 - az)(1 - bz)}{(1 - cz)(1 - dz)} \to A(z + \frac{k}{z}) + B. \qquad (5.66)$$

Normalizing $A = 1, B = 0$, this gives q-quadratic grid: $x_s = q^s + kq^{-s}$, which is utilized in this chapter. The interpolation on q-grid $x_s = q^s$ is a degeneration $k \to 0$ of this. There are further degenerations to $x_s = s^2 + as$ (additive quadratic grid), $x_s = s$ (additive grid [36, 39]), and also the differential grid (usual Padé approximation).

Various applications of these Padé problems to the (discrete) Painlevé equations can be found in the references cited below (the corresponding chapters in this book are also indicated).

Summary of the results

We summarize the results obtained in this and previous chapters.

- **Padé approximation (differential grid)** (Chapters 2 and 3)

$$\psi(x) = (1 - x)^\kappa \prod_{i=1}^{N} (1 - \frac{x}{t_i})^{\alpha_i} : \mathrm{Gar}_N \ (\dim = 2N) \to P_{\mathrm{VI}} \ (N = 1) \to \text{(a)}$$

$\uparrow$ continuous limit $q \to 1$

$$\psi(x) = \prod_{i=1}^{N+1} \frac{(a_i x)_\infty}{(b_i x)_\infty} : q\text{-}\mathrm{Gar}_N \ (\dim = 2N) \to q\text{-}P_{\mathrm{VI}} \ (N = 1) \to \text{(b)}$$

$\downarrow$ constraint $q^m \prod_{i=1}^{N+1} a_i = q^n \prod_{i=1}^{N+1} b_i$

$\dim = 2(N - 1) \to q\text{-}E_6^{(1)} \ (N = 2) \to \text{(b)}$

- **Padé interpolation on q-grid** (Chapter 4)

$$\psi(x) \propto x^\gamma \prod_{i=1}^{N} \frac{(a_i x)_\infty}{(b_i x)_\infty} \to q\text{-}\mathrm{Gar}_N \to q\text{-}P_{\mathrm{VI}} \ (N = 1) \to \text{(c)}$$

$\downarrow$ constraint $q^m \prod_{i=1}^{N} a_i = q^n \prod_{i=1}^{N} b_i$ or $\gamma = 0$

$\dim = 2(N - 1) \to q\text{-}E_6^{(1)} \ (N = 2) \to \text{(c)}$

$\downarrow$ constraint $q^m \prod_{i=1}^{N} a_i = q^n \prod_{i=1}^{N} b_i$ and $\gamma = 0$

$\dim = 2(N - 2) \to q\text{-}E_7^{(1)} \ (N = 3) \to \text{(c)}$

- **Padé interpolation on q-quadratic grid** (Chapter 5)

$$\psi(x) \propto \prod_{i=1}^{N} \frac{(a_i x, \frac{qka_i}{x})_\infty}{(b_i x, \frac{qkb_i}{x})_\infty} \rightarrow \dim = 2(N-1) \rightarrow q\text{-}E_7^{(1)} \ (N=2)$$

$$\downarrow \quad \text{constraint } q^m \prod_{i=1}^{N} a_i = q^n \prod_{i=1}^{N} b_i$$

$$\dim = 2(N-2) \rightarrow q\text{-}E_8^{(1)} \ (N=3)$$

For further degenerate cases, see the following references (in the tables below, "HGF in sol." means the hypergeometric functions appearing in the special solutions):

(a) [63], where the following $\psi(x)$ are studied:

	P_{VI}	P_V	P_{IV}
$\psi(x)$	$(1-x)^a \left(1 - \dfrac{x}{t}\right)^b$	$(1-x)^{-b} \exp\left(\dfrac{xt}{1-x}\right)$	$\exp(2tx - x^2)$
HGF in sol.	Gauss $_2F_1$	Laguerre $_1F_1$	Hermite $_1F_1$

(b) [37], where the following $\psi(x)$ are studied:

	$q\text{-}E_6^{(1)}$	$q\text{-}D_5^{(1)} \ (q\text{-}P_{VI})$	$q\text{-}A_4^{(1)}$	$q\text{-}(A_2 + A_1)^{(1)}$
$\psi(x)$	$\prod_{i=1}^{3} \dfrac{(a_i x)_\infty}{(b_i x)_\infty}$ $\dfrac{a_1 a_2 a_3 q^m}{b_1 b_2 b_3 q^n} = 1$	$\prod_{i=1}^{2} \dfrac{(a_i x)_\infty}{(b_i x)_\infty}$	$\dfrac{(a_1 x, a_2 x)_\infty}{(b_1 x)_\infty}$	$(a_1 x, a_2 x)_\infty$
HGF in sol.	$_3\varphi_2$	$_2\varphi_1$	$_2\varphi_1$	$_1\varphi_1$

(c) [35], where the following $\psi(x)$ are studied:

	$q\text{-}E_7^{(1)}$	$q\text{-}E_6^{(1)}$	$q\text{-}D_5^{(1)} \ (q\text{-}P_{VI})$	$q\text{-}A_4^{(1)}$	$q\text{-}(A_2 + A_1)^{(1)}$
$\psi(x)$	$\prod_{i=1}^{3} \dfrac{(a_i x, b_i)_\infty}{(a_i, b_i x)_\infty}$ $\dfrac{a_1 a_2 a_3 q^m}{b_1 b_2 b_3 q^n} = 1$	$\prod_{i=1}^{2} \dfrac{(a_i x, b_i)_\infty}{(a_i, b_i x)_\infty}$	$c^{\log_q x} \dfrac{(a_1 x, b_1)_\infty}{(a_1, b_1 x)_\infty}$	$c^{\log_q x} \dfrac{(b_1)_\infty}{(b_1 x)_\infty}$	$(d\sqrt{x/q})^{\log_q x}$
ψ_s	$\prod_{i=1}^{3} \dfrac{(b_i)_s}{(a_i)_s}$	$\prod_{i=1}^{2} \dfrac{(b_i)_s}{(a_i)_s}$	$c^s \dfrac{(b_1)_s}{(a_1)_s}$	$c^s (b_1)_s$	$q^{s(s-1)/2} d^s$
HGF in sol.	$_4\varphi_3$	$_3\varphi_2$	$_2\varphi_1$	$_2\varphi_1$	$_1\varphi_1$

Chapter 6
Multicomponent Generalizations

Abstract We will discuss generalizations of isomonodromic deformations to higher order or higher rank, using the multicomponent Padé approximations. Two kinds of approximation/interpolation problems and their duality relations play an important role.

6.1 Padé Approximations With Multicomponent

Let ℓ be a positive integer and let $\mathbf{f}$ be an ℓ-tuple of analytic functions around $x = 0$,

$$\mathbf{f} = (f_1, f_2, \ldots, f_\ell). \tag{6.1}$$

Following [2] (see also [30]) we consider two kinds of approximation problems (HP) and (SP). We put $\mathbf{n} = (n_1, n_2, \ldots, n_\ell) \in \mathbb{Z}_{\geq 0}^\ell$ and $|\mathbf{n}| = \sum_{i=1}^\ell n_i$.

(HP) The *Hermite–Padé problem* is a problem to find polynomials $Q_i^{(\mathbf{n})}(x)$ $(i = 1, \ldots, \ell)$ such that

$$\deg Q_i^{(\mathbf{n})}(x) \leq n_i - 1 \quad (i = 1, \ldots, \ell), \tag{6.2}$$

$$\sum_{i=1}^\ell f_i(x) Q_i^{(\mathbf{n})}(x) \equiv 0 \pmod{x^{|\mathbf{n}|-1}}. \tag{6.3}$$

The number of free coefficients is $|\mathbf{n}|$ and the number of equations is $|\mathbf{n}| - 1$, hence the polynomials $Q_1^{(\mathbf{n})}(x), \ldots, Q_\ell^{(\mathbf{n})}(x)$ can be uniquely determined up to a common normalization.

© The Author(s), under exclusive license to Springer Nature Singapore Pte Ltd 2021
H. Nagao and Y. Yamada, *Padé Methods for Painlevé Equations*,
SpringerBriefs in Mathematical Physics,
https://doi.org/10.1007/978-981-16-2998-3_6

(SP) The *simultaneous Padé problem* is a problem to find polynomials $P_i^{(\mathbf{n})}(x)$ $(i = 1, \ldots, \ell)$ such that

$$\deg P_i^{(\mathbf{n})}(x) \leq |\mathbf{n}| - n_i \quad (i = 1, \ldots, \ell),$$

$$\frac{P_1^{(\mathbf{n})}(x)}{f_1(x)} \equiv \cdots \equiv \frac{P_\ell^{(\mathbf{n})}(x)}{f_\ell(x)} \pmod{x^{|\mathbf{n}|+1}}. \tag{6.4}$$

The number of free coefficients is $\sum_{i=1}^{\ell}(|\mathbf{n}| - n_i + 1) = (\ell - 1)|\mathbf{n}| + \ell$, and the number of equations is $(\ell - 1)(|\mathbf{n}| + 1)$, hence the polynomials $P_1^{(\mathbf{n})}(x), \ldots, P_\ell^{(\mathbf{n})}(x)$ can be uniquely determined up to an overall normalization.

In some sense, the above two problems are dual to each other. To see this, define $\ell \times \ell$ matrices $Q(x)$ and $\mathcal{P}(x)$ with entries

$$Q_{ij}(x) = Q_j^{(\mathbf{n}+\mathbf{e}_i)}(x), \quad \mathcal{P}_{ij}(x) = P_j^{(\mathbf{n}-\mathbf{e}_i)}(x) \quad (i, j = 1, 2 \ldots, \ell). \tag{6.5}$$

We normalize them so that the diagonal entries are monic for convenience.

Theorem 6.1 (Mahler duality[2]) *The following relation holds:*

$$Q(x)\mathcal{P}(x)^T = x^{|\mathbf{n}|} I, \tag{6.6}$$

and $|Q(x)| = x^{|\mathbf{n}|}$, $|\mathcal{P}(x)| = x^{(\ell-1)|\mathbf{n}|}$.

Proof Consider the (i, j)-element of the LHS: $M_{ij} = \sum_{k=1}^{\ell} Q_k^{(\mathbf{n}+\mathbf{e}_i)} P_k^{(\mathbf{n}-\mathbf{e}_j)}$. We see that

$$\deg M_{ij} \leq \sum_{k=1}^{\ell}\{\deg Q_k^{(\mathbf{n}+\mathbf{e}_i)} + \deg P_k^{(\mathbf{n}-\mathbf{e}_j)}\} \tag{6.7}$$

$$\leq \sum_{k=1}^{\ell}\{(n_k + \delta_{ik} - 1) + (|\mathbf{n}| - 1 - n_k + \delta_{jk})\}$$

$$\leq |\mathbf{n}| - 1 + \delta_{ij},$$

and the diagonal elements M_{ii} are monic of degree $|\mathbf{n}|$. On the other hand, by the conditions (6.2) and (6.4), we have

$$M_{ij} = \sum_{k=1}^{\ell} f_k Q_k^{(\mathbf{n}+\mathbf{e}_i)} \frac{P_k^{(\mathbf{n}-\mathbf{e}_j)}}{f_k} \equiv 0 \pmod{x^{|\mathbf{n}|-1}}, \tag{6.8}$$

hence the equation (6.6) is proved. From $\deg|Q(x)| \leq |\mathbf{n}|$, $\deg|\mathcal{P}(x)| \leq (\ell - 1)|\mathbf{n}|$ and (6.6), we have $|Q(x)| = x^{|\mathbf{n}|}$, $|\mathcal{P}(x)| = x^{(\ell-1)|\mathbf{n}|}$. $\qquad\square$

The importance of the Mahler duality in the isomonodromic deformation theory was noted by T. Mano and T. Tsuda [30, 31], where a very interesting application of the duality to the *Schlesinger transformation* was given. As the simplest case, we will construct some special solutions of the (differential) Garnier system in the next section.

6.2 Application to the *N*-Garnier System

Here, we will construct some special solutions of the Garnier systems using the Padé method [29] (see also [63]). Consider a system of linear differential equations

$$\partial_x \Psi(x) = \mathcal{A}(x)\Psi(x), \quad \partial_{t_i}\Psi(x) = \mathcal{B}_i(x)\Psi(x), \tag{6.9}$$

$$\mathcal{A}(x) = \sum_{i=0}^{N+1} \frac{M_i}{x - t_i}, \quad \mathcal{B}_i(x) = -\frac{M_i}{x - t_i}, \tag{6.10}$$

where M_i ($i = 0, \ldots, N + 1$) are $\ell \times \ell$ matrices depending on $t = (t_0, \ldots, t_N)$ but independent of x. Their compatibility is given by the *Schlesinger system*

$$\partial_{t_i} M_j = \frac{[M_i, M_j]}{t_i - t_j} \ (i \neq j), \quad \partial_{t_i} M_i = - \sum_{j(\neq i)=0}^{N} \frac{[M_j, M_i]}{t_j - t_i}, \tag{6.11}$$

which can be written as a *multi-time Hamiltonian system* $\partial_{t_i} f = \{H_i, f\}$ with the following Hamiltonians H_i and Poisson bracket:

$$\mathrm{tr}\mathcal{A}(x)^2 = \sum_{i=0}^{N+1} \left\{ \frac{\alpha^2}{(x - t_i)^2} + 2\frac{H_i}{x - t_i} \right\}, \tag{6.12}$$

$$H_i = \sum_{j(\neq i)=0}^{N+1} \frac{\mathrm{tr}(M_i M_j)}{t_i - t_j}, \tag{6.13}$$

$$\{(M_i)_{ab}, (M_j)_{cd}\} = \delta_{ij}\Big(\delta_{bc}(M_i)_{ad} - \delta_{ad}(M_i)_{cb}\Big). \tag{6.14}$$

It is known that $H_1, \ldots, H_{N+1}$ are a Poisson limit of the Gaudin Hamiltonians [11] (see also [6]), and the Schlesinger system can be considered as a classical limit of the Knizhnik–Zamolodchikov equation (see [13] for example).

In the case of $\ell = 2$, assuming the eigenvalues of M_i to be $\{0, \alpha_i\}$, we can parametrize the matrices M_i as

$$M_i = \begin{bmatrix} 1 \\ q_i \end{bmatrix} [a_i - q_i p_i, \ p_i] = \begin{bmatrix} \alpha_i - q_i p_i & p_i \\ q_i(a_i - q_i p_i) & q_i p_i \end{bmatrix}, \tag{6.15}$$

in terms of canonical coordinates q_i, p_i such that $\{p_i, q_j\} = \delta_{ij}$. We have three conserved quantities ($M_\infty := -\sum_{i=0}^{N+1} M_i$) and one rescaling freedom[1] $q_i \to \lambda q_i$, $p_i \to \lambda^{-1} p_i$, hence, four variables among $2(N+2)$ are redundant. The resulting system with $2N$ variables is equivalent to the N-Garnier system. It is important that the Hamiltonians are polynomials in q_i, p_i variables [26] but not so in λ_i, μ_i. The coordinates $\{(\lambda_i, \mu_i)\}$ are given by the "magic recipe" [54]: $\mathcal{A}(\lambda_i)_{12} = 0$ and $\mu_i = \mathcal{A}(\lambda_i)_{11}$.

We will derive a class of special solutions of the N-Garnier system using the Padé method. To do this, we consider the Hermite–Padé problem[2] with $\ell = 2$, $\mathbf{n} = (m, n)$, namely

$$Q_{i1}(x)f_1(x) + Q_{i2}(x)f_2(x) = O(x^{m+n}) \quad (i = 1, 2), \tag{6.16}$$

where $Q_{ij}(x)$ are polynomials such that

$$Q(x) = \begin{bmatrix} Q_{11}(x) & Q_{12}(x) \\ Q_{21}(x) & Q_{22}(x) \end{bmatrix}, \quad \deg Q(x) = \begin{bmatrix} m & n-1 \\ m-1 & n \end{bmatrix}, \tag{6.17}$$

and $Q_{11}(x)$, $Q_{22}(x)$ are monic.

Lemma 6.1 *We have the following explicit expressions of $Q_{ij}(x)$ in terms of the Schur functions associated with $\psi(x)$ (6.23):*

$$Q_{11}(x) = x^m \frac{V^*(x)\tau_{m,n}}{\tau_{m,n}}, \tag{6.18}$$

$$Q_{12}(x) = (-1)^n x^{n-1} \frac{V(x)\tau_{m+1,n-1}}{\tau_{m,n}}, \tag{6.19}$$

$$Q_{21}(x) = (-1)^{n-1} x^{m-1} \frac{V^*(x)\tau_{m-1,n+1}}{\tau_{m,n}}, \tag{6.20}$$

$$Q_{22}(x) = x^n \frac{V(x)\tau_{m,n}}{\tau_{m,n}}, \tag{6.21}$$

where V and V^* are the operators defined in Proposition 2.3.

Proof The Padé problem (6.16) is equivalent to a pair of scalar Padé approximations with [numerator/denominator] polynomials of degree $[m/(n-1)]$ and $[(m-1)/n]$ respectively:

$$\psi(x) = \frac{Q_{11}(x)}{Q_{12}(x)} + O(x^{m+n}), \quad \psi(x) = \frac{Q_{21}(x)}{Q_{22}(x)} + O(x^{m+n}), \tag{6.22}$$

where

$$\psi(x) = -\frac{f_2(x)}{f_1(x)}. \tag{6.23}$$

---

[1] For $\mathcal{A}(x)$, $\mathcal{B}(x)$, this rescaling is realized as a gauge transformation by a diagonal matrix.

[2] The (HP) and (SP) are equivalent for $\ell = 2$.

Then, from the results in Sect. 2.5, we have the desired formulas.   □

We put

$$\Psi(x) = Q(x).d_f(x), \quad d_f(x) = \text{diag}(f_1(x), f_2(x)). \tag{6.24}$$

Then, the matrix $\Psi(x)$ gives a fundamental solution of the matrix Lax pair of the isomonodromic type (6.9), where

$$\mathcal{A}(x) = \partial_x(Q(x)d_f(x))d_f(x)^{-1}Q(x)^{-1}, \tag{6.25}$$
$$\mathcal{B}_i(x) = \partial_{t_i}(Q(x)d_f(x))d_f(x)^{-1}Q(x)^{-1}. \tag{6.26}$$

We will consider the case where

$$f_1(x) = -1, \quad f_2(x) = \psi(x) = \prod_{i=1}^{N+1}(1 - \frac{x}{t_i})^{\alpha_i}. \tag{6.27}$$

Lemma 6.2 *The matrices $\mathcal{A}(x), \mathcal{B}(x)$ can be written as*

$$\mathcal{A}(x) = \sum_{i=0}^{N+1}\frac{M_i}{x - t_i}, \quad \mathcal{B}_i(x) = -\frac{M_i}{x - t_i} - \begin{bmatrix} 0 & 0 \\ 0 & \alpha_i \end{bmatrix}\frac{1}{t_i}, \tag{6.28}$$

where M_i are as in (6.15).

Proof We note that $\mathcal{A}(x), \mathcal{B}(x)$ are gauge transformations of

$$(\partial_x d_f)d_f^{-1} = \sum_{i=1}^{N+1}\begin{bmatrix} 0 & 0 \\ 0 & \alpha_i \end{bmatrix}\frac{1}{x - t_i}, \tag{6.29}$$

$$(\partial_{t_i}d_f)d_f^{-1} = -\begin{bmatrix} 0 & 0 \\ 0 & \alpha_i \end{bmatrix}(\frac{1}{x - t_i} + \frac{1}{t_i}), \tag{6.30}$$

by $Q(x)$. Moreover, we have $|Q(x)| = x^{|\mathbf{n}|}$ and $Q(x) = (I + O(\frac{1}{x}))\text{diag}(x^m, x^n)$. Hence the singularities of $\mathcal{A}(x), \mathcal{B}(x)$ are only simple poles at $x = t_0(= 0), t_1, \ldots, t_{N+1}, t_\infty(= \infty)$, and the x-dependence of $\mathcal{A}(x), \mathcal{B}(x)$ is as in (6.28). The eigenvalues of M_i are: $(0, |\mathbf{n}|)$ for M_0, $(0, \alpha_i)$ for M_i $(1 \le i \le N + 1)$ and $(m, n + \sum_{i=1}^{N+1}\alpha_i)$ for $-M_\infty = \sum_{i=0}^{N+1}M_i$. Hence, the variables q_i, p_i can be introduced as in (6.15). □

Theorem 6.2 *Define $q_i(t), p_i(t)$ $(i = 1, 2, \ldots, N + 1)$ as*

$$q_i = (-1)^{n-1}t_i T_{\alpha_i}\left(\frac{\tau_{m,n}}{\tau_{m+1,n-1}}\right), \tag{6.31}$$

$$q_i - \frac{\alpha_i}{p_i} = \frac{(-1)^n}{t_i}T_{\alpha_i}^{-1}\left(\frac{\tau_{m-1,n+1}}{\tau_{m,n}}\right), \tag{6.32}$$

where $T_{\alpha_i} : \alpha_i \to \alpha_i + 1$. And further define $q_0(t)$, $p_0(t)$ by the constraints

$$\sum_{i=0}^{N+1} p_i = 0, \quad \sum_{i=0}^{N+1} (\alpha_i - q_i p_i) = m. \tag{6.33}$$

Then these functions satisfy the multi-time Hamiltonian system with Hamiltonian

$$H_i = \sum_{j(\neq i)=0}^{N+1} \frac{\mathrm{tr}(M_i M_j)}{t_i - t_j} - \frac{\alpha_i}{t_i} \sum_{j=0}^{N+1} q_j p_j. \tag{6.34}$$

Proof By construction, $\mathcal{A}(x)$, $\mathcal{B}(x)$ (the on-Padé situation) are compatible and give a special solution of the N-Garnier system. The explicit forms of solutions and equations can be determined as follows. The formulae of $\{q_i, p_i\}$ are obtained by looking at the left and right kernel of the rank one matrices $M_i = \mathrm{Res}_{x=t_i}\mathcal{A}(x)$ (6.15). Then, from Lemmas 6.1 and 6.2, we have

$$q_i = \frac{Q_{22}(t_i)}{Q_{12}(t_i)} = (-1)^{n-1} t_i T_{\alpha_i}\left(\frac{\tau_{m,n}}{\tau_{m+1,n-1}}\right), \tag{6.35}$$

$$q_i - \frac{\alpha_i}{p_i} = \frac{Q_{21}(t_i)}{Q_{22}(t_i)} = \frac{(-1)^n}{t_i} T_{\alpha_i}^{-1}\left(\frac{\tau_{m-1,n+1}}{\tau_{m,n}}\right), \tag{6.36}$$

$(i = 1, 2, \ldots, N + 1)$. Here we have used the fact that the action of the operators $V(x)$, $V^*(x)$ for $x = t_i$ can be rewritten by the shifts T_{α_i}, $T_{\alpha_i}^{-1}$ thanks to the form of $\psi(x)$ (6.27). The constraints (6.33) follow from $M_\infty = \mathrm{diag}(m, n + \sum_{i=1}^{N+1} \alpha_i)$.

To determine the equation explicitly, we note that the matrices (6.28) can be obtained from (6.9) by a gauge transformation $\mathcal{A}(x) \to g\mathcal{A}(x)g^{-1} + (\partial_x g)g^{-1}$, $\mathcal{B}(x) \to g\mathcal{B}(x)g^{-1} + (\partial_{t_i} g)g^{-1}$, with $g = \mathrm{diag}(1, \prod_{i=1}^{N+1} t_i^{-\alpha_i})$. Correspondingly, the canonical variables (q_i, p_i) are transformed as

$$q_i \to t_i^{-\alpha_i} q_i, \quad p_i \to t_i^{\alpha_i} p_i, \quad H_i \to H_i - \sum_{i=0}^{N+1} \frac{\alpha_i}{t_i} q_i p_i. \tag{6.37}$$

As a result, the q_i, p_i in (6.31) satisfy the Hamiltonian system with the Hamiltonian (6.34). $\square$

6.3 Discrete Mahler Duality

Consider the following discrete analog of (6.2):

(HP) The *discrete Hermite–Padé problem*

$$\deg Q_i^{(\mathbf{n})}(x) \leq n_i - 1 \quad (i = 1, \ldots, \ell),$$

$$\sum_{i=1}^{\ell} f_i(x) Q_i^{(\mathbf{n})}(x) = 0 \quad (\text{for } x = x_s, \ s = 0, 1, \ldots, |\mathbf{n}| - 2). \tag{6.38}$$

Similarly, a discrete analog of (6.4) can be formulated as

(SP) The *discrete simultaneous Padé problem*

$$\deg P_i^{(\mathbf{n})}(x) \leq |\mathbf{n}| - n_i \quad (i = 1, \ldots, \ell),$$

$$\frac{P_1^{(\mathbf{n})}(x)}{f_1(x)} = \cdots = \frac{P_\ell^{(\mathbf{n})}(x)}{f_\ell(x)} \quad (\text{for } x = x_s, \ s = 0, 1, \ldots, |\mathbf{n}|). \tag{6.39}$$

Define $\ell \times \ell$ matrices Q and $\mathcal{P}$ with entries

$$Q_{i,j} = Q_j^{(\mathbf{n}+\mathbf{e}_i)}, \quad \mathcal{P}_{i,j} = P_j^{(\mathbf{n}-\mathbf{e}_i)} \quad (i, j = 1, 2 \ldots, \ell). \tag{6.40}$$

We normalize them so that the diagonal entries are monic for convenience.

Proposition 6.1 (discrete Mahler duality) *The following relation holds:*

$$Q(x)\mathcal{P}^T(x) = \prod_{s=0}^{|\mathbf{n}|-1} (x - x_s)I. \tag{6.41}$$

Proof The proof is almost the same as the differential case. Namely, first we note $\deg M_{ij} \leq |\mathbf{n}| - 1 + \delta_{ij}$, and by the conditions (6.38), (6.39) we have

$$M_{ij} = \sum_{k=1}^{\ell} f_k Q_k^{(\mathbf{n}+\mathbf{e}_i)} \frac{P_k^{(\mathbf{n}-\mathbf{e}_j)}}{f_k} = 0, \tag{6.42}$$

for $x = x_s$ ($s = 0, 1, \ldots, |\mathbf{n}| - 1$). $\qquad\square$

Triangular gauge

The elements of $Q(x)$, $\mathcal{P}(x)$ are polynomials such that $\deg Q_{i,j}(x) = n_j - 1 + \delta_{ij}$ and $\deg \mathcal{P}_{i,j}(x) = |\mathbf{n}| - n_j - 1 + \delta_{ij}$. In the following we will modify the matrices to the following triangular form:

$Q(x) = \text{(upper triangular)} + O(x) \quad (x \to 0)$,
$Q(x)\text{diag}(x^{-n_1}, \dots, x^{-n_\ell}) = I + \text{(strict lower triangular)} + O(\frac{1}{x}) \quad (x \to \infty)$,
$\mathcal{P}(x)^T = \text{(upper triangular)} + O(x) \quad (x \to 0)$,
$x^{-|\mathbf{n}|}\text{diag}(x^{n_1}, \dots, x^{n_\ell})\mathcal{P}(x)^T = I + \text{(strict lower triangular)} + O(\frac{1}{x}) \quad (x \to \infty)$,
$$\tag{6.43}$$

which is realized by a conjugation by a constant lower triangular matrix. In this gauge, the matrices $Q(x)$, $\mathcal{P}(x)$ still have the duality relation (6.41).

As an application, consider the q-grid case $x_s = q^s$ with the functions $\mathbf{f}(x) = (f_1(x), \dots, f_\ell(x))$ given by

$$f_i(x) = c_i^{\log_q x} \prod_{k=1}^{m_i} \frac{(b_i x)_\infty}{(a_i x)_\infty}, \tag{6.44}$$

for generic parameters a_i, b_i, c_i. Then, in the triangular gauge, the matrix

$$Y(x) = Q(x)\text{diag}(\mathbf{f}(x)) \tag{6.45}$$

satisfy the difference equation

$$Y(qx) = \mathcal{A}(x)Y(x), \tag{6.46}$$

where $\mathcal{A}(x)$ has the following properties:

$$\mathcal{A}(x) = \frac{\text{polynomial in } x \text{ of degree } (1 + \sum_{i=1}^\ell m_i)}{(x - q^{|\mathbf{n}|-1}) \prod_{i=1}^\ell \prod_{k=1}^{m_i}(1 - b_i x)}, \tag{6.47}$$

$$\det \mathcal{A}(x) = \frac{(1 - qx)}{(1 - xq^{1-|\mathbf{n}|})} \prod_{i=1}^\ell \left(c_i \prod_{k=1}^{m_i} \frac{1 - a_i x}{1 - b_i x}\right), \tag{6.48}$$

$$\lim_{x \to 0} \mathcal{A}(x) \text{ is upper triangular with diagonals } c_i, \tag{6.49}$$

$$\lim_{x \to \infty} \mathcal{A}(x) \text{ is lower triangular with diagonals } c_i q^{n_i} \prod_{k=1}^{m_i} \frac{a_k}{b_k}. \tag{6.50}$$

And it can be written in factorized form as

$$\mathcal{A}(x) = \text{diag}(*)(\underbrace{X(*)X(*)\cdots X(*)}_{N})^{-1}(\underbrace{X(*)X(*)\cdots X(*)}_{N}), \tag{6.51}$$

where $N = 1 + \sum_{i=1}^\ell m_i$ and $X(*)$ is the matrix of the following form

$$X(u) = \mathrm{diag}(u) + \begin{bmatrix} 0 & 1 & & & \\ & 0 & 1 & & \\ & & \ddots & \ddots & \\ & & & 0 & 1 \\ x & & & & 0 \end{bmatrix}, \quad u \in \mathbb{C}^\ell. \tag{6.52}$$

The isomonodromic deformations of the equation (6.46) with $\mathcal{A}(x)$ (6.51) are generalizations of the system considered in [24]. The case $\ell = 2$ gives a realization of the q-Garnier system by factorized Lax matrices as in [46, 49].

Remark 6.1 The matrices of the form $X(*)$ and $X(*)^{-1}$ play fundamental roles in the theory of soliton cellular automata with reflecting ends [27].

References

1. Ablowitz, M.J., Clarkson, P.A.: Solitons, Nonlinear Evolution Equations and Inverse Scattering. Lecture Notes in Mathematics, vol. 149. Cambridge University Press, Cambridge (1991)
2. Baker, G., Graves-Morris, P.: Padé Approximants, 2nd edn. Cambridge University Press, Cambridge (1996)
3. Cauchy, A.L.: Cours d'Analyse de l'Ecole Royale Polytechnique. Première Partie. Analyse algébrique. Imprimérie Royale, Paris (1821)
4. Chudnovsky, D.V., Chudnovsky, G.V.: Bäcklund transformations for linear differential equations and Padé approximations I. J. Math. Pures Appl. **61**, 1–16 (1982)
5. Date, E.: On a direct method of constructing multi-soliton solutions. Proc. Jpn. Acad. **15**, Ser. A, 27–29 (1979)
6. Feigin, B., Frenkel, E., Reshetikhin, N.: Gaudin model, Bethe ansatz and critical level. Commun. Math. Phys. **166**, 27–62 (1994)
7. Fuchs, R.: Sur quelques équations différentielles linéaires du second ordre. C. R. Acad. Sci. (Paris) **141**, 555–558 (1905)
8. Garnier, R.: Sur des équations différentielles du troisiéme ordre dont l'intégrale générale est uniforme et sur unc classe d'équations nouvelles d'ordre supérieur dont l'intégrale générale a ses points critiques fixes. Ann. Sci. Ecole Norm. Super. **29**, 1–126 (1912)
9. Garnier, R.: Etudes de l'intégrale générale de l'équation VI de M. Painlevé dans le voisinage de ses singularités transcendentes. Ann. Sci. Ecole Norm. Super. **34**(3), 239–353 (1917)
10. Gasper, G., Rahman, M.: Basic Hypergeometric Series. With a foreword by Richard Askey, 2nd edn. Encyclopedia of Mathematics and Its Applications, vol. 96. Cambridge University Press, Cambridge (2004)
11. Gaudin, M.: Diagonalisation dúne classe dhamiltoniens de spin. J. Phys. Fr. **37**, 1087–1098 (1976)
12. Grammaticos, B., Ramani, A.: On a novel q-discrete analogue of the Painlevé VI equation. Phys. Lett. A **257**, 288–292 (1999)
13. Harnad, J.: Quantum isomonodromic deformations and the Knizhnik-Zamolodchikov equations. CRM Proc. Lect. Notes **9**, 155–161 (Am. Math. Soc., Providence, RI, 1996)
14. Hietarinta, J., Joshi, N., Nijhoff, F.W.: Discrete Systems and Integrability. Cambridge Texts in Applied Mathematics. Cambridge University Press (2016)
15. Hirota, R.: Discrete analogue of a generalized Toda equation. J. Phys. Soc. Jpn. **50**(11), 3785–3791 (1981)

16. Ikawa, Y.: Hypergeometric solutions for the q-Painlevé equation of type $E_6^{(1)}$ by the Padé method. Lett. Math. Phys. **103**, 743–763 (2013)
17. Iwasaki, K., Kimura, H., Shimomura, S., Yoshida, M.: From Gauss to Painlevé: A Modern Theory of Special Functions. Vieweg, Braunschweig (1991)
18. Jacobi, C.G.J.: Über die Darstellung einer Reihe gegebner Werthe durch eine gebrochne rationale function. J. Reine Angew. Math. **30**, 127–156 (1846)
19. Jimbo, M., Sakai, H.: A q-analog of the sixth Painlevé equation. Lett. Math. Phys. **38**, 145–154 (1996)
20. Jimbo, M., Miwa, T., Ueno, K.: Monodromy preserving deformation of linear ordinary differential equations with rational coefficients I. Physica D.**2**, 306–52 (1981)
21. Jimbo, M., Miwa, T.: Monodromy preserving deformation of linear ordinary differential equations with rational coefficients II. Physica D.**2**, 407–448 (1981)
22. Jimbo, M., Miwa, T.: Monodromy preserving deformation of linear ordinary differential equations with rational coefficients III. Physica D.**4**, 26–46 (1981)
23. Kajihara, Y., Noumi, M.: Multiple elliptic hypergeometric series - an approach from Cauchy determinant. Indag. Math. New Ser. **14**, 395–421 (2003)
24. Kajiwara, K., Noumi, M., Yamada, Y.: q-Painlevé systems arising from q-KP hierarchy. Lett. Math. Phys. **62**(3), 259–268 (2002)
25. Kajiwara, K., Noumi, M., Yamada, Y.: Geometric aspects of Painlevé equations. J. Phys. A: Math. Theor. **50**(073001), 164 (2017) (Topical Review)
26. Kimura, H., Okamoto, K.: On the polynomial Hamiltonian structure of the Garnier systems. J. Math. Pures Appl. **63**, 129–146 (1984)
27. Kuniba, A., Okado, M., Yamada, Y.: Box-ball system with reflecting end. J. Nonlinear Math. Phys. **12**, 475–507 (2005)
28. Magnus, A.: Painlevé-type differential equations for the recurrence coefficients of semiclassical orthogonal polynomials. J. Comput. Appl. Math. **57**, 215–237 (1995)
29. Mano, T.: Determinant formula for solutions of the Garnier system and Padé approximation. J. Phys. A: Math. Theor. **45**, 135206–135219 (2012)
30. Mano, T., Tsuda, T.: Two approximation problems by Hermite and the Schlesinger transformations. (Japanese). RIMS Kokyuroku Bessatsu **B47**, 77–86 (2014)
31. Mano, T., Tsuda, T.: Hermite-Padé approximation, isomonodromic deformation and hypergeometric integral. Math. Z. **285**, 397–431 (2017)
32. Masuda, T.: On a class of algebraic solutions to the Painlevé VI equation, its determinant formula and coalescence cascade. Funkc. Ekvacioj **46**, 121–171 (2003)
33. Masuda, T.: Hypergeometric τ-functions of the q-Painlevé system of type $E_8^{(1)}$. Ramanujan J. **24**, 1–31 (2011)
34. Miwa, T., Jimbo, M., Date, E.: Solitons: differential equations symmetries and infinite dimensional algebras. Cambridge Tracts in Mathematics, vol. 135. Cambridge University Press (2000)
35. Nagao, H.: The Padé interpolation method applied to q-Painlevé equations. Lett. Math. Phys. **105**, 503–521 (2015)
36. Nagao, H.: Lax pairs for additive difference Painlevé equations. arXiv:1604.02530 [nlin.SI]
37. Nagao, H.: The Padé interpolation method applied to q-Painlevé equations II (differential grid version). Lett. Math. Phys. **107**, 107–127 (2017)
38. Nagao, H.: A variation of the q-Painlevé system with affine Weyl group symmetry of type $E_7^{(1)}$. SIGMA **13**(092), 18 (2017)
39. Nagao, H.: The Padé interpolation method applied to additive difference Painlevé equations. arXiv:1706.10101 [nlin.SI]
40. Nagao, H., Yamada, Y.: Variations of the q-Garnier system. J. Phys. A: Math. Theor. **51**, 135204–135222 (2018)
41. Nagao, H., Yamada, Y.: Study of q-Garnier system by Padé method. Funkc. Ekvacioj **61**, 109–133 (2018)
42. Noumi, M., Tsujimoto, S., Yamada, Y.: Padé interpolation for elliptic Painlevé equation. In: Symmetries, Integrable Systems and Representations. Springer Proceedings in Mathematics and Statistics, vol. 40, pp. 463–482. Springer (2013)

43. Noumi, M., Ruijsenaars, S., Yamada, Y.: The elliptic Painlevé Lax equation vs. van Diejen's 8-coupling elliptic Hamiltonian. SIGMA **16**(063), 16 (2020)
44. Ohta, Y., Ramani, A., Grammaticos, B.: An affine Weyl group approach to the eight-parameter discrete Painlevé equation. J. Phys. A **34**, 10523–10532 (2001)
45. Okamoto, K.: Studies on the Painlevé equations. I. Sixth Painlevé equation P_{VI}. Ann. Mat. Pura Appl. **146**(4), 337–381 (1987)
46. Ormerod, C.M., Rains, E.M.: Commutation relations and discrete Garnier systems. SIGMA **12**(110), 50 (2016)
47. Ormerod, C.M., Rains, E.M.: An elliptic Garnier system. Commun. Math. Phys. **355**, 741–766 (2017)
48. Padé, H.: Sur l'expression générale de la fonction rationnelle approchée de $(1 + x)^m$. C. R. Acad. Sci. Paris **132**, 754–756 (1901)
49. Park, K.: A certain generalization of q-hypergeometric functions and their related monodromy preserving deformation. J. Integr. Syst. **3**, xyy019 (2018)
50. Ramani, A., Grammaticos, B., Hietarinta, J.: Discrete versions of the Painlevé equations. Phys. Rev. Lett. **67**, 1829–1832 (1991)
51. Sakai, H.: Rational surfaces with affine root systems and geometry of the Painlevé equations. Commun. Math. Phys. **220**, 165–221 (2001)
52. Sakai, H.: A q-analog of the Garnier system. Funkc. Ekvacioj **48**, 273–297 (2005)
53. Sakai, H.: Lax form of the q-Painlevé equation associated with the $A_2^{(1)}$ surface. J. Phys. A: Math. Gen. **39**, 12203–12210 (2006)
54. Sklyanin, E.K.: Separation of variables - new trends. In: Quantum Field Theory, Integrable Models and Beyond. (Kyoto, 1994). Prog. Theor. Phys. Suppl. **118**, 35–60 (1995)
55. Sklyanin, E.K., Takebe, T.: Separation of variables in the elliptic Gaudin model. Commun. Math. Phys. **204**, 17–38 (1999)
56. Suzuki, T.: A reformulation of the generalized q-Painlevé VI system with $W(A_{2n+1}^{(1)})$ symmetry. J. Integr. Syst. **2**, 1–18 (2017)
57. Spiridonov, V.P.: Essays on the theory of elliptic hypergeometric functions. Russ. Math. Surv. **63**, 405–472 (2008)
58. Takemura, K.: Degenerations of Ruijsenaars-van Diejen operator and q-Painlevé equations. J. Integr. Syst. **2**(1), xyx008, 27 (2017)
59. Tsuda, T.: Rational solutions of the Garnier system in terms of Schur polynomials. Int. Math. Res. Not. **2003**, 2341–58 (2003)
60. Tsuda, T.: On an integrable system of q-difference equations satisfied by the universal characters: its Lax formalism and an application to q-Painlevé equations. Commun. Math. Phys. **293**, 347–359 (2010)
61. Tsuda, T.: Hypergeometric solution of a certain polynomial Hamiltonian system of isomonodromy type. Q. J. Math. **63**, 489–505 (2012)
62. Van Assche, W.: Orthogonal Polynomials and Painlevé Equations. Australian Mathematical Society Lecture Series, vol. 27. Cambridge University Press, Cambridge (2018)
63. Yamada, Y.: Padé method to Painlevé equations. Funkc. Ekvacioj **52**, 83–92 (2009)
64. Yamada, Y.: Lax formalism for q-Painlevé equations with affine Weyl group symmetry of type $E_n^{(1)}$. IMRN **17**, 3823–3838 (2011)
65. Yamada, Y.: A simple expression for discrete Painlevé equations. RIMS Kokyuroku Bessatsu **B47**, 087–095 (2014)
66. Yamada, Y.: An elliptic Garnier system from interpolation. SIGMA **13**(069), 8 (2017)
67. Whittaker, E.T., Watson, G.N.: A Course of Modern Analysis, 4th edn, p. 608. Cambridge University Press (1928)
68. Zhedanov, A.S.: Padé interpolation table and biorthogonal rational functions. In Elliptic Integrable Systems. Rokko Lectures in Mathematics, vol. 18, pp. 323–363. Kobe University (2005)

Index

© The Author(s), under exclusive license to Springer Nature Singapore Pte Ltd 2021
H. Nagao and Y. Yamada, *Padé Methods for Painlevé Equations*,
SpringerBriefs in Mathematical Physics,
https://doi.org/10.1007/978-981-16-2998-3

Printed in the United States
by Baker & Taylor Publisher Services